SpringerBriefs in Optimization

SpringerBriefs in Optimization showcases algorithmic and theoretical techniques, case studies, and applications within the broad-based field of optimization. Manuscripts related to the ever-growing applications of optimization in applied mathematics, engineering, medicine, economics, and other applied sciences are encouraged.

More information about this series at http://www.springer.com/series/8918

Alexander J. Zaslavski

Optimal Control Problems Arising in Forest Management

 Springer

Alexander J. Zaslavski
Department of Mathematics
The Technion – Israel Institute of Techn
Rishon LeZion, Israel

ISSN 2190-8354   ISSN 2191-575X   (electronic)
SpringerBriefs in Optimization
ISBN 978-3-030-23586-4   ISBN 978-3-030-23587-1   (eBook)
https://doi.org/10.1007/978-3-030-23587-1

This Springer imprint is published by the registered company Springer Nature Switzerland AG.
The registered company address is: Gewerbestrasse 11, 6330 Cham, Switzerland

Preface

The growing importance of optimal control has been recognized in recent years. This is due not only to impressive theoretical developments but also because of numerous applications to engineering, economics, life sciences, etc. This book is devoted to the study of a class of optimal control problems arising in forest management. The forest management problem is an important and interesting topic in mathematical economics that was studied by many researchers including Nobel laureate P.A. Samuelson [68]. As usual, for this problem, the existence of optimal solutions over infinite horizon and the structure of solutions on finite intervals are under consideration. In our books [84, 86], we study a class of discrete-time optimal control problems which describe many models of economic dynamics except for the model of forest management. This happens because some assumptions posed in [84, 86], which are true for many models of economic dynamics, do not hold for the model of forest management. By this reason, the forest management problem is not a particular case of general models of economic dynamics and is studied separately in the literature. In this book, we study the forest management problem using the approach introduced and employed in our research [80, 81, 83]. Namely, we analyze a class of optimal control problems which contains, as a particular case, the forest management problem. For this class of problems, we show the existence of optimal solutions over infinite horizon and study the structure of approximate solutions on finite intervals and their turnpike properties, the stability of the turnpike phenomenon, and the structure of approximate solutions on finite intervals in the regions close to the endpoints.

In Chap. 1, we provide some preliminary knowledge on turnpike properties. The forest management problem is discussed in Chap. 2, which also contains existence results for infinite horizon problems. In Chap. 3, we establish the turnpike properties of approximate solutions. Chapter 4 contains generic turnpike results. We consider a class of optimal control problems which is identified with a complete metric space of objective functions and show the existence of a G_δ everywhere dense subset of the metric space, which is a countable intersection of open everywhere dense sets, such that the turnpike property holds for any of its element. Chapter 5 is devoted to the study of the structure of approximate solutions on finite intervals in the regions close to the endpoints. In Chap. 6, we again consider the forest management problem and show that the results of Chaps. 3 and 5 are true for it.

Rishon LeZion, Israel   Alexander J. Zaslavski
October 30, 2018

Contents

Chapter 1
Introduction

The study of optimal control problems and variational problems defined on infinite intervals and on sufficiently large intervals has been a rapidly growing area of research [9, 10, 16, 18, 24, 37, 43, 48, 57, 59, 66, 76, 84–86] which has various applications in engineering [1, 44, 87], in models of economic growth [2, 5, 14, 20, 23, 28, 38–40, 47, 52–55, 60, 61, 63, 65, 67, 68, 70, 77, 80, 81, 83], in the game theory [29, 32, 42, 74, 82], in infinite discrete models of solid-state physics related to dislocations in one-dimensional crystals [6, 71], and in the theory of thermodynamical equilibrium for materials [19, 45, 50, 51]. Discrete-time problems were considered in [7, 8, 13, 21, 26, 30, 33, 72, 73, 78, 79] while continuous-time problems were studied in [3, 4, 11, 12, 15, 17, 22, 25, 27, 31, 41, 46, 49, 56, 58, 62, 69, 75].

In this chapter we discuss turnpike properties for a class of simple convex dynamic optimization problems.

1.1 Convex Discrete-Time Problems

Let R^n be the n-dimensional Euclidean space with the inner product $\langle \cdot, \cdot \rangle$ which induces the norm

$$|x| = \left(\sum_{i=1}^{n} x_i^2 \right)^{1/2}, \ x = (x_1, \ldots, x_n) \in R^n.$$

Let $v : R^n \times R^n \rightarrow R^1$ be bounded from below function. We consider the minimization problem

© The Author(s), under exclusive license to Springer Nature Switzerland AG 2019
A. J. Zaslavski, *Optimal Control Problems Arising in Forest Management*,
SpringerBriefs in Optimization, https://doi.org/10.1007/978-3-030-23587-1_1

$$\sum_{i=0}^{T-1} v(x_i, x_{i+1}) \to \min, \qquad\qquad (P_0)$$

such that $\{x_i\}_{i=0}^{T} \subset R^n$ and $x_0 = z$, $x_T = y$,

where T is a natural number and the points $y, z \in R^n$.

The interest in discrete-time optimal problems of type (P_0) stems from the study of various optimization problems which can be reduced to it, e.g., continuous-time control systems which are represented by ordinary differential equations whose cost integrand contains a discounting factor [43], tracking problems in engineering [1, 44, 87], the study of Frenkel-Kontorova model [6, 71], and the analysis of a long slender bar of a polymeric material under tension in [19, 45, 50, 51]. Optimization problems of the type (P_0) were considered in [72, 73].

In this section we suppose that the function $v : R^n \times R^n \to R^1$ is strictly convex and differentiable and satisfies the growth condition

$$v(y, z)/(|y| + |z|) \to \infty \text{ as } |y| + |z| \to \infty. \qquad\qquad (1.1)$$

We intend to study the behavior of solutions of the problem (P_0) when the points y, z and the real number T vary and T is sufficiently large. Namely, we are interested to study a turnpike property of solutions of (P_0) which is independent of the length of the interval T, for all sufficiently large intervals. To have this property means, roughly speaking, that solutions of the optimal control problems are determined mainly by the objective function v and are essentially independent of T, y, and z. Turnpike properties are well known in mathematical economics. The term was first coined by Samuelson in 1948 (see [67]) where he showed that an efficient expanding economy would spend most of the time in the vicinity of a balanced equilibrium path (also called von Neumann path). This property was further investigated for optimal trajectories of models of economic dynamics (see, for example, [47, 53, 65] and the references mentioned there). Many turnpike results are collected in [76, 84, 86].

In order to meet our goal we consider the auxiliary optimization problem

$$v(x, x) \to \min, \ x \in R^n. \qquad\qquad (P_1)$$

It follows from the strict convexity of v and (1.1) that the problem (P_1) has a unique solution $\bar{x}$. Let

$$\nabla v(\bar{x}, \bar{x}) = (l_1, l_2), \qquad\qquad (1.2)$$

where $l_1, l_2 \in R^n$. Since $\bar{x}$ is a solution of (P_1) it follows from (1.2) that for each $h \in R^n$,

$$\langle l_1, h \rangle + \langle l_2, h \rangle = \langle (l_1, l_2), (h, h) \rangle$$
$$= \lim_{t \to 0^+} t^{-1}[v(\bar{x} + th, \bar{x} + th) - v(\bar{x}, \bar{x})] \geq 0.$$

Thus

$$\langle l_1 + l_2, h \rangle \geq 0 \text{ for all } h \in R^n,$$

$l_2 = -l_1$ and

$$\nabla v(\bar{x}, \bar{x}) = (l_1, -l_1), \tag{1.3}$$

For each $(y, z) \in R^n \times R^n$ set

$$L(y, z) = v(y, z) - v(\bar{x}, \bar{x}) - \langle \nabla v(\bar{x}, \bar{x}), (y - \bar{x}, z - \bar{x}) \rangle$$
$$= v(y, z) - v(\bar{x}, \bar{x}) - \langle l_1, y - z \rangle. \tag{1.4}$$

It is not difficult to verify that the function $L : R^n \times R^n \to R^1$ is differentiable and strictly convex. It follows from (1.1) and (1.4) that

$$L(y, z)/(|y| + |z|) \to \infty \text{ as } |y| + |z| \to \infty. \tag{1.5}$$

Since the functions v and L are both strictly convex it follows from (1.4) that

$$L(y, z) \geq 0 \text{ for all } (y, z) \in R^n \times R^n \tag{1.6}$$

and

$$L(y, z) = 0 \text{ if and only if } y = \bar{x}, \ z = \bar{x}. \tag{1.7}$$

We claim that the function $L : R^n \times R^n \to R^1$ has the following property:
(C) If a sequence $\{(y_i, z_i)\}_{i=1}^{\infty} \subset R^n \times R^n$ satisfies the equality

$$\lim_{i \to \infty} L(y_i, z_i) = 0,$$

then

$$\lim_{i \to \infty} (y_i, z_i) = (\bar{x}, \bar{x}).$$

Assume that a sequence $\{(y_i, z_i)\}_{i=1}^{\infty} \subset R^n \times R^n$ satisfies $\lim_{i \to \infty} L(y_i, z_i) = 0$. In view of (1.5) the sequence $\{(y_i, z_i)\}_{i=1}^{\infty}$ is bounded. Let (y, z) be its limit point. Then it is easy to see that the equality

$$L(y, z) = \lim_{i \to \infty} L(y_i, z_i) = 0$$

holds and by (1.7) $(y, z) = (\bar{x}, \bar{x})$. This implies that $(\bar{x}, \bar{x}) = \lim_{i \to \infty} (y_i, z_i)$.
Thus the property (C) holds, as claimed.

Consider an auxiliary minimization problem

$$\sum_{i=0}^{T-1} L(x_i, x_{i+1}) \rightarrow \min, \tag{P_2}$$

such that $\{x_i\}_{i=0}^{T} \subset R^n$ and $x_0 = z$, $x_T = y$,

where T is a natural number and the points $y, z \in R^n$.

It follows from (1.4) that for any integer $T \geq 1$ and any sequence $\{x_i\}_{i=0}^{T} \subset R^n$, we have

$$\sum_{i=0}^{T-1} L(x_i, x_{i+1}) = \sum_{i=0}^{T-1} v(x_i, x_{i+1}) - Tv(\bar{x}, \bar{x}) - \sum_{i=0}^{T-1} \langle l_1, x_i - x_{i+1}\rangle$$

$$= \sum_{i=0}^{T-1} v(x_i, x_{i+1}) - Tv(\bar{x}, \bar{x}) - \langle l_1, x_0 - x_T\rangle. \tag{1.8}$$

Relation (1.8) implies that the problems (P_0) and (P_2) are equivalent. Namely, $\{x_i\}_{i=0}^{T} \subset R^n$ is a solution of the problem (P_0) if and only if it is a solution of the problem (P_2).

Let T be a natural number and $\Delta \geq 0$. A sequence $\{x_i\}_{i=0}^{T} \subset R^n$ is called (Δ)-optimal if for any sequence $\{x_i'\}_{i=0}^{T} \subset R^n$ satisfying $x_i = x_i'$, $i = 0, T$ the inequality

$$\sum_{i=0}^{T-1} v(x_i, x_{i+1}) \leq \sum_{i=0}^{T-1} v(x_i', x_{i+1}') + \Delta$$

holds. Clearly, if a sequence $\{x_i\}_{i=0}^{T} \subset R^n$ is (0)-optimal, then it is a solution of the problems (P_0) and (P_2) with $z = x_0$ and $y = x_T$.

We prove the following existence result.

Proposition 1.1 *Let $T > 1$ be an integer and $y, z \in R^n$. Then the problem (P_0) has a solution.*

Proof It is sufficient to show that the problem (P_2) has a solution. Consider a sequence $\{x_i'\}_{i=0}^{T} \subset R^n$ such that $x_0' = z$, $x_T' = y$. Set

$$M_1 = \sum_{i=0}^{T-1} L(x_i', x_{i+1}')$$

and

$$M_2 = \inf \left\{ \sum_{i=0}^{T-1} L(x_i, x_{i+1}) : \{x_i\}_{i=0}^{T} \subset R^n, \ x_0 = z, \ x_T = y \right\}. \tag{1.9}$$

Clearly,

$$0 \le M_2 \le M_1.$$

We may assume without loss of generality that

$$M_2 < M_1. \tag{1.10}$$

There exists a sequence $\{x_i^{(k)}\}_{i=0}^T \subset R^n$, $k = 1, 2, \ldots$ such that for any natural number k,

$$x_0^{(k)} = z, \quad x_T^{(k)} = y \tag{1.11}$$

and

$$\lim_{k \to \infty} \sum_{i=0}^{T-1} L(x_i^{(k)}, x_{i+1}^{(k)}) = M_2. \tag{1.12}$$

In view of (1.10), (1.11), and (1.12) we may assume that

$$\sum_{i=0}^{T-1} L(x_i^{(k)}, x_{i+1}^{(k)}) < M_1 \text{ for all integers } k \ge 1. \tag{1.13}$$

By (1.13) and (1.5) there is $M_3 > 0$ such that

$$|x_i^{(k)}| \le M_3 \text{ for all } i = 0, \ldots, T \text{ and all integers } k \ge 1. \tag{1.14}$$

In view of (1.14), extracting subsequences, using diagonalization process and re-indexing, if necessary, we may assume without loss of generality that for each $i \in \{0, \ldots, T\}$ there exists

$$\widehat{x}_i = \lim_{k \to \infty} x_i^{(k)}. \tag{1.15}$$

By (1.15) and (1.11),

$$\widehat{x}_0 = z, \quad \widehat{x}_T = y. \tag{1.16}$$

It follows from (1.15) and (1.12) that

$$\sum_{i=0}^{T-1} L(\widehat{x}_i, \widehat{x}_{i+1}) = M_2.$$

Together with (1.16) and (1.9) this implies that $\{\widehat{x}_i\}_{i=0}^T$ is a solution of the problem (P_2). This completes the proof of Proposition 1.1.

Denote by Card (A) the cardinality of a set A.

The following result establishes a turnpike property for approximate solutions of the problem (P_0).

Proposition 1.2 *Let M_1, M_2, ϵ be positive numbers. Then there exists a natural number k_0 such that for each integer $T > 1$ and each (M_1)-optimal sequence $\{x_i\}_{i=0}^T \subset R^n$ satisfying*

$$|x_0| \leq M_2, \ |x_T| \leq M_2, \tag{1.17}$$

the following inequality holds:

$$Card(\{i \in \{0, \ldots, T-1\} : |x_i - \bar{x}| + |x_{i+1} - \bar{x}| > \epsilon\}) \leq k_0.$$

Proof By condition (C) there is $\delta > 0$ such that for each $(y, z) \in R^n \times R^n$ satisfying

$$L(y, z) \leq \delta, \tag{1.18}$$

we have

$$|y - \bar{x}| + |z - \bar{x}| \leq \epsilon. \tag{1.19}$$

Set

$$M_3 = \sup\{L(y, z) : \ y, z \in R^n \text{ and } |y| + |z| \leq |\bar{x}| + M_2\}. \tag{1.20}$$

Choose a natural number

$$k_0 > \delta^{-1}(M_1 + 2M_3). \tag{1.21}$$

Assume that an integer $T > 1$ and that an (M_1)-optimal sequence $\{x_i\}_{i=0}^T \subset R^n$ satisfies (1.17). Set

$$y_0 = x_0, \ y_T = x_T, \ y_i = \bar{x}, \ i = 1, \ldots, T-1. \tag{1.22}$$

Since the sequence $\{x_i\}_{i=0}^T$ is (M_1)-optimal it follows from (1.22) that

$$\sum_{i=0}^{T-1} v(x_i, x_{i+1}) \leq \sum_{i=0}^{T-1} v(y_i, y_{i+1}) + M_1.$$

Together with (1.7), (1.8), and (1.22) this implies that

$$\sum_{i=0}^{T-1} L(x_i, x_{i+1}) \le \sum_{i=0}^{T-1} L(y_i, y_{i+1}) + M_1 = L(x_0, \bar{x}) + L(\bar{x}, x_T) + M_1.$$

Combined with (1.17) and (1.20) this implies that

$$\sum_{i=0}^{T-1} L(x_i, x_{i+1}) \le M_1 + 2M_3.$$

It follows from the choice of δ (see (1.18) and (1.19)), (1.21), and the inequality above that

$$\mathrm{Card}(\{i \in \{0, \ldots, T-1\} : |x_i - \bar{x}| + |x_{i+1} - \bar{x}| > \epsilon\})$$

$$\le \mathrm{Card}(\{i \in \{0, \ldots, T-1\} : L(x_i, x_{i+1}) > \delta\})$$

$$\le \delta^{-1} \sum_{i=0}^{T-1} L(x_i, x_{i+1}) \le \delta^{-1}(M_1 + 2M_3) \le k_0.$$

Proposition 1.2 is proved.

Proposition 1.2 implies the following turnpike result for exact solutions of the problem (P_0).

Proposition 1.3 *Let M, ϵ be positive numbers. Then there exists a natural number k_0 such that for each integer $T > 1$, each $y, z \in R^n$ satisfying $|y|, |z| \le M$, and each optimal solution $\{x_i\}_{i=0}^{T} \subset R^n$ of the problem (P_0) the following inequality holds:*

$$\mathrm{Card}(\{i \in \{0, \ldots, T-1\} : |x_i - \bar{x}| + |x_{i+1} - \bar{x}| > \epsilon\}) \le k_0.$$

It is easy now to see that the optimal solution $\{x_i\}_{i=0}^{T}$ of the problem (P_0) spends most of the time in an ϵ-neighborhood of $\bar{x}$. By Proposition 1.3 the number of all integers $i \in \{0, \ldots, T-1\}$, such that x_i does not belong to this ϵ-neighborhood, does not exceed the constant k_0 which depends only on M, ϵ and does not depend on T. Following the tradition, the point $\bar{x}$ is called the turnpike. Moreover we can show that the set

$$\{i \in \{0 \ldots, T\} : |x_i - \bar{x}| > \epsilon\}$$

is contained in the union of two intervals $[0, k_1] \cup [T - k_1, T]$, where k_1 is a constant depending only on M, ϵ.

1.2   The Turnpike Phenomenon

In the previous section we proved the turnpike result for rather simple class of
discrete-time problems. The problems of this class are unconstrained and their
objective functions are convex and differentiable. It should be mentioned that,
using the methods of convex analysis [64], this result can be extended to the class
of objective functions which are merely convex. In the turnpike theory and its
applications to the mathematical economics our goal is to establish the turnpike
property for constrained optimal control problems. In particular, in this book and in
[84, 86] we study the structure of approximate solutions of an autonomous discrete-
time control system with a compact metric space of states X. This control system
is described by a nonempty closed set $\Omega \subset X \times X$ which determines a class of
admissible trajectories (programs) and a bounded upper semicontinuous function
$v : \Omega \to R^1$ which determines an optimality criterion. We study the problem

$$\sum_{i=0}^{T-1} v(x_i, x_{i+1}) \to \max, \ \{(x_i, x_{i+1})\}_{i=0}^{T-1} \subset \Omega, \ x_0 = z, \ x_T = y, \qquad (P)$$

where $T \geq 1$ is an integer and the points $y, z \in X$. Clearly, these constrained
problems are more difficult and less understood than their unconstrained prototypes.
They are also more realistic from the point of view of mathematical economics.

In the turnpike theory the objective function v possesses the turnpike property
(TP) if there exists a point $\bar{x} \in X$ (a turnpike) such that the following condition
holds:

For each positive number ϵ there exists an integer $L \geq 1$ such that for each
integer $T \geq 2L$ and each solution $\{x_i\}_{i=0}^T \subset X$ of the problem (P) the inequality
$\rho(x_i, \bar{x}) \leq \epsilon$ is true for all $i = L, \ldots, T - L$.

It should be mentioned that the constant L depends neither on T nor on y, z.

The turnpike phenomenon has the following interpretation. If one wishes to reach
a point A from a point B by a car in an optimal way, then one should turn to a
turnpike, spend most of the time on it, and then leave the turnpike to reach the
required point.

It should be mentioned that in general a turnpike is not necessarily a singleton
[76]. Nevertheless problems of the type (P) for which the turnpike is a singleton
are of great importance because of the following reasons: there are many models of
economic growth for which a turnpike is a singleton; if a turnpike is a singleton, then
approximate solutions of (P) have very simple structure and this is very important
for applications; if a turnpike is a singleton, then it can be easily calculated as a
solution of the problem $v(x, x) \to \max, (x, x) \in \Omega$.

The turnpike property is very important for applications. Suppose that our
objective function v has the turnpike property and we know a finite number of
"approximate" solutions of the problem (P). Then we know the turnpike $\bar{x}$, or at
least its approximation, and the constant L (see the definition of (TP)) which is an
estimate for the time period required to reach the turnpike. This information can

be useful if we need to find an "approximate" solution of the problem (P) with a new time interval $[m_1, m_2]$ and the new values $z, y \in X$ at the end points m_1 and m_2. Namely instead of solving this new problem on the "large" interval $[m_1, m_2]$ we can find an "approximate" solution of the problem (P) on the "small" interval $[m_1, m_1 + L]$ with the values $z, \bar{x}$ at the end points and an "approximate" solution of the problem (P) on the "small" interval $[m_2 - L, m_2]$ with the values $\bar{x}, y$ at the end points. Then the concatenation of the first solution, the constant sequence $x_i = \bar{x}, i = m_1 + L, \ldots, m_2 - L$ and the second solution is an "approximate" solution of the problem (P) on the interval $[m_1, m_2]$ with the values z, y at the end points. Sometimes as an "approximate" solution of the problem (P) we can choose any admissible sequence $\{x_i\}_{i=m_1}^{m_2}$ satisfying

$$x_{m_1} = z, \ x_{m_2} = y \text{ and } x_i = \bar{x} \text{ for all } i = m_1 + L, \ldots, m_2 - L.$$

In this book we study a class of optimal control problems arising in the forest management. The forest management problem is an important and interesting topic in mathematical economics studies by many researchers. We consider the discrete time model for the optimal management of a forest of total area S occupied by k species $I = \{1, \ldots, k\}$ with maturity ages of $n_1, \ldots, n_k$ years respectively. This model and its versions were studied in [20, 38–40, 54, 55, 61, 68, 80, 81, 83]. Here the optimal harvesting of a mixed forest composed of multiple species is studied, each one having a different maturity age, where only mature trees can be harvested.

As usual, for this problem the existence of optimal solutions over infinite horizon and the structure of solutions on finite intervals are under consideration. In our books [84, 86] we study a class of discrete-time optimal control problems which describe many models of economic dynamics except of the model of the forest management. This happens because some assumptions posed in [84, 86], which are true for many models of economic dynamics, do not hold for the model of the forest management. Namely, in [84, 86] we studied optimal control problems of type (P) for which the turnpike property holds with the turnpike $\bar{x}$ such that $(\bar{x}, \bar{x})$ is an interior point of the set $\Omega \subset X \times X$. This assumption, which plays an important role in [84, 86], holds for many models of economic dynamics except of the forest management problem considered in this book.

By this reason, the forest management problem is not a particular case of general models of economic dynamics and is studied separately in the literature. In this book we study the forest management problem using the approach introduced and employed in our research [80, 81, 83]. Namely, we analyze a class of optimal control problems which contains, as a particular case, the forest management problem. For this class of problems we show the existence of optimal solutions over infinite horizon, study the structure of approximate solutions on finite intervals and their turnpike properties, the stability of the turnpike phenomenon, and the structure of approximate solutions on finite intervals in the regions close to the end points.

Chapter 2
Infinite Horizon Optimal Control Problems

In this chapter we present the forest management problem and study some properties of its solutions. This problem is a particular case of a general optimal control problem which is also introduced in this chapter. The corresponding infinite horizon problems without discounting as well as with discounting are considered. The results of the chapter were obtained from [80, 81].

2.1 The Forest Management Problem

We consider a discrete time model for the optimal management of a forest of total area S occupied by k species $I = \{1, \ldots, k\}$ with maturity ages of $n_1, \ldots, n_k$ years respectively. This model and its versions were studied in [20, 38–40, 54, 55, 61, 68, 80, 81, 83]. Here the optimal harvesting of a mixed forest composed of multiple species is studied, each one having a different maturity age, where only mature trees can be harvested.

For each period $t = 0, 1, \ldots$ we denote $x_i^j(t) \geq 0$ the area covered by trees of species i that are j years old with $j = 1, \ldots, n_i$ and $\bar{x}_i(t) \geq 0$ the area occupied by over-mature trees (older than n_i). We must decide how much land $v_i(t) \geq 0$ to harvest and how to reallocate this land to new seedlings.

Assuming that only mature trees can be harvested we must have

$$v_i(t) \leq \bar{x}_i(t) + x_i^{n_i}(t), \tag{2.1}$$

and then the area not harvested in that period will comprise the over-mature trees at the next step, namely

$$\bar{x}_i(t+1) = \bar{x}_i(t) + x_i^{n_i}(t) - v_i(t). \tag{2.2}$$

© The Author(s), under exclusive license to Springer Nature Switzerland AG 2019
A. J. Zaslavski, *Optimal Control Problems Arising in Forest Management*,
SpringerBriefs in Optimization, https://doi.org/10.1007/978-3-030-23587-1_2

The fact that immature trees cannot be harvested is represented by

$$x_i^{j+1}(t+1) = x_i^j(t), \quad j = 1, \ldots, n_i - 1. \tag{2.3}$$

The total harvested area $\sum_{i \in I} v_i(t)$ is allocated to new seedlings which is expressed by the equation

$$\sum_{i \in I} x_i^1(t+1) = \sum_{i \in I} v_i(t). \tag{2.4}$$

In the sequel we use the notation

$$x_i^{n_i+1} = \bar{x}_i, \quad i \in I. \tag{2.5}$$

A representation of the forest in terms of the age distribution at time t is provided by the state $x(t) = (x_1(t), \ldots, x_k(t))$, where

$$x_i(t) = (x_i^1(t), \ldots, x_i^{n_i}(t), x_i^{n_i+1}(t)), \quad i \in I$$

describes the areas occupied in year t by trees of species i with ages $1, 2, \ldots, n_i$ and over n_j. The first and last components of each vector $x_i(t)$ are controlled by the sowing and harvesting policies. Note that we do not control $x(0)$ which corresponds to the initial state reflecting the age class composition of the forest at time $t = 0$.

Let $R_+^m = \{x = (x_1, \ldots, x_m) \in R^m : x_i \geq 0, \ i = 1, \ldots, m\}$ and let $N = \sum_{i \in I}(n_i + 1)$. Every vector $x \in R^N$ is represented as $x = (x_1, \ldots, x_k)$, where $x_i = (x_i^1, \ldots, x_i^{n_i}, x_i^{n_i+1}) \in R^{n_i+1}$ for all integers $i = 1, \ldots, k$.

Denote by Δ the set of all $x \in R_+^N$ such that

$$\sum_{i \in I} \left[\sum_{j=1}^{n_i+1} x_i^j \right] = S. \tag{2.6}$$

Now we give a formal description of the model.

A sequence $\{x(t)\}_{t=0}^\infty \subset \Delta$ is called a program if for all integers $t \geq 0$ and all $i \in I$ Eqs. (2.1)–(2.4) hold (see (2.5)) with some $v(t) = (v_1(t), \ldots, v_k(t)) \in R_+^k$.

Let the integers T_1, T_2 satisfy $0 \leq T_1 < T_2$. A sequence $\{x(t)\}_{t=T_1}^{T_2} \subset \Delta$ is called a program if Eqs. (2.1)–(2.4) hold for all $i \in I$ and for all integers $t = T_1, \ldots, T_2-1$ (see (2.5)) with some $v(t) = (v_1(t), \ldots, v_k(t)) \in R_+^k$.

An alternative equivalent definition of a program can be given with the help of the transition possibility. Put

$$\Omega = \left\{ (x, y) \in \Delta \times \Delta : y_i^{j+1} = x_i^j \text{ for all } i \in I \text{ and all } j \in \{1, \ldots, n_i\} \setminus \{n_i\} \right.$$

$$\left. \text{and for all } i \in I, \ x_i^{n_i+1} + x_i^{n_i} - y_i^{n_i+1} \geq 0 \right\}. \tag{2.7}$$

Clearly, if $(x, y) \in \Omega$, then

$$\sum_{i \in I} y_i^1 = \sum_{i \in I} (x_i^{n(i)+1} + x_i^{n_i} - y_i^{n(i)+1}). \tag{2.8}$$

It is easy to see that a sequence $\{x(t)\}_{t=0}^{\infty} \subset \Delta$ is a program if and only if $(x(t), x(t+1)) \in \Omega$ for all integers $t \geq 0$.

Let the integers T_1, T_2 satisfy $0 \leq T_1 < T_2$. It is easy to see that a sequence $\{x(t)\}_{t=T_1}^{T_2} \subset \Delta$ is a program if and only if $(x(t), x(t+1)) \in \Omega$ for all $t = T_1, \ldots, T_2 - 1$.

For each $(x, y) \in \Omega$ set

$$V(x, y) = (v_1(x, y), \ldots, v_k(x, y)),$$

where for $i = 1, \ldots, k$,

$$v_i(x, y) = x_i^{n_i+1} + x_i^{n_i} - y_i^{n_i+1}.$$

Define

$$\Delta_0 = \left\{ v \in R_+^k : \sum_{i=1}^{k} v_i \leq S \right\}.$$

In this book we assume that a benefit at moment $t = 0, 1, \ldots$ is represented by an upper semicontinuous function $w_t : \Delta_0 \to R^1$ and at a moment $t = 0, 1, \ldots,$ $w_t(V(x, y))$ is the benefit obtained today if the forest today is x and the forest tomorrow is y, where $(x, y) \in \Omega$.

Note that usually in the literature it is assumed that for $t = 0, 1, \ldots,$

$$w_t(V(x, y)) = \sum_{i=1}^{k} W^{(i)}(v_i(x, y)), \quad (x, y) \in \Omega$$

where $W^i : [0, \infty) \to R^1, i = 1, \ldots, k$ are strictly concave, smooth, and increasing functions.

Clearly, Δ is a compact set in R^N, Ω is a closed subset of $\Delta \times \Delta$, and $w_t \circ V : \Omega \to R^1$ is an upper semicontinuous function for all integers $t \geq 0$. Set

$$\bar{n} = \max\{n_i : i \in I\}. \tag{2.9}$$

Our model has an important property established by the following result. Its proof is given in [80].

Proposition 2.1 *Let* $x, y \in \Delta$. *Then there exists a program* $\{x(t)\}_{t=0}^{N+\bar{n}+1}$ *such that* $x(0) = x$ *and* $x(N + \bar{n} + 1) = y$.

Proof Put $x(0) = x$. For all integers $t = 0, \ldots, N - 1$ define

$$x_i^{j+1}(t + 1) = x_i^j(t), \ i \in I, \ j \in \{1, \ldots, n_i\} \setminus \{n_i\},$$

$$x_i^1(t + 1) = 0, \ i \in I,$$

$$x_i^{n_i+1}(t + 1) = x_i^{n_i+1}(t) + x_i^{n_i}(t), \ i \in I. \tag{2.10}$$

It is easy to see that $x(t) \in \Delta$ for all $t = 0, \ldots, N$, $\{x(t)\}_{t=0}^N$ is a program and

$$x_i^j(N) = 0, \ i \in I, \ j = 1, \ldots, n_i,$$

$$\sum_{i \in I} x_i^{n_i+1}(N) = S. \tag{2.11}$$

For each $s = 1, \ldots, \bar{n}$ put

$$I_s = \{i \in I : \ n_i = s\}. \tag{2.12}$$

(Note that for some integers s we can have $I_s = \emptyset$.)
    We assume that sum over empty set is zero.
    Define $x(N + 1) \in \Delta$ as follows. Set

$$x_i^1(N + 1) = y_i^{n_i+1}, \ i \in I_{\bar{n}}, \tag{2.13}$$

$$x_i^1(N + 1) = 0, \ i \in I \setminus I_{\bar{n}}. $$

For $i \in I$ and all integers j satisfying $1 < j \leq n_i$ set

$$x_i^j(N + 1) = 0. \tag{2.14}$$

Clearly, there exist

$$u_i \in [0, x_i^{n_i+1}(N)], \ i \in I \tag{2.15}$$

such that

$$\sum_{i \in I} u_i = \sum_{i \in I_{\bar{n}}} x_i^1(N + 1). \tag{2.16}$$

Put

$$x_i^{n_i+1}(N + 1) = x_i^{n_i+1}(N) - u_i, \ i \in I. \tag{2.17}$$

By (2.13)–(2.17), $x(N + 1) \in \Delta$ and

$$(x(N), x(N + 1)) \in \Omega.$$

Assume that q is an integer, $1 \leq q < \bar{n}$, and we have defined a program $\{x(t)\}_{t=0}^{N+q}$ such that the following properties hold:

(P1)   If an integer

$$i \in \cup\{I_s : \quad \text{an integer } s \text{ satisfies } 1 \leq s \leq \bar{n} - q\},$$

then $x_i^j(N + q) = 0$ for all integers j satisfying $1 \leq j \leq n_i$;

(P2)   If an integer s satisfies $\bar{n} \geq s > \bar{n} - q$ and $i \in I_s$, then

$$x_i^p(N + q) = y_i^{n_i+1+p-(q+s-\bar{n})}, \quad p = 1, \ldots, q + s - \bar{n}, \tag{2.18}$$

$$x_i^p(N+q) = 0 \text{ for all integers } p \text{ satisfying } q+s-\bar{n} < p \leq n_i. \tag{2.19}$$

(Note that for $q = 1$ our assumptions hold.)

Define $x(N+q+1) \in \Delta$ as follows. Let $i \in I$. If $i \in I_s$, where $1 \leq s \leq \bar{n}-q-1$, then set

$$x_i^j(N + q + 1) = 0 \text{ for all integers } j \text{ satisfying } 1 \leq j \leq n_i. \tag{2.20}$$

If $i \in I_{\bar{n}-q}$, then set

$$x_i^1(N + q + 1) = y_i^{n_i+1}, \quad x_i^p(N + q + 1) = 0 \tag{2.21}$$

for all integers p satisfying $1 < p \leq n_i$.

If $i \in I_s$, where $\bar{n} \geq s > \bar{n} - q$, then set (see (2.18))

$$x_i^{p+1}(N + q + 1) = x_i^p(N + q) = y_i^{n_i+1+p-(q+s-\bar{n})}, \quad p = 1, \ldots, q + s - \bar{n}, \tag{2.22}$$

$$x_i^1(N + q + 1) = y_i^{n_i+1-(q+s-\bar{n})}, \tag{2.23}$$

$$x_i^p(N+q+1) = 0 \text{ for all integers } p \text{ satisfying } q+1+s-\bar{n} < p \leq n_i. \tag{2.24}$$

It is not difficult to see that

$$\sum_{i \in I} x_i^1(N + q + 1) \leq \sum_{i \in I} x_i^{n_i+1}(N + q). \tag{2.25}$$

Therefore there exists

$$x_i^{n_i+1}(N+q+1) \in \left[0, x_i^{n_i+1}(N+q)\right], \quad i \in I, \tag{2.26}$$

such that

$$\sum_{i \in I}[x_i^{n_i+1}(N+q) - x_i^{n_i+1}(N+q+1)] = \sum_{i \in I} x_i^1(N+q+1). \tag{2.27}$$

Clearly,

$$x(N+q+1) \in \Delta, \quad (x(N+q), x(N+q+1)) \in \Omega$$

and the assumption made for q holds also for $q+1$. Thus by induction we have constructed a program $\{x(t)\}_{t=0}^{N+\bar{n}}$ such that (P1) and (P2) hold for all $q = 1, \ldots, \bar{n}$.
Consider the state $x(N+\bar{n})$. Let $i \in I_s$ where $1 \leq s \leq \bar{n}$. By (P1) and (P2),

$$x_i^p(N+\bar{n}) = y_i^{p+1}, \quad p = 1, \ldots, n_i. \tag{2.28}$$

Set

$$x(N+\bar{n}+1) = y. \tag{2.29}$$

By (2.28) and (2.29),

$$(x(N+\bar{n}), x(N+\bar{n}+1)) \in \Omega.$$

Proposition 2.1 is proved.

2.2 Infinite Horizon Problems Without Discounting

Let (K, ρ) be a compact metric space and Ω be a nonempty closed subset of $K \times K$.
  A sequence $\{x_t\}_{t=0}^{\infty} \subset K$ is called an (Ω)-program (or a program if the set Ω is understood) if $(x_t, x_{t+1}) \in \Omega$ for all integers $t \geq 0$.
  Let the integers T_1, T_2 satisfy $T_1 < T_2$. A sequence $\{x_t\}_{t=T_1}^{T_2} \subset K$ is called an (Ω)-program (or a program if the set Ω is understood) if $(x_t, x_{t+1}) \in \Omega$ for all integers t satisfying $T_1 \leq t < T_2$.
  For each integer $t \geq 0$, let $u_t : \Omega \to R^1$ be a bounded upper semicontinuous function.
  For each pair of integers T_1, T_2 satisfying $0 \leq T_1 < T_2$ and each $y, z \in K$ we consider the optimization problems

$$\sum_{i=T_1}^{T_2-1} u_i(x_i, x_{i+1}) \to \max,$$

$$\{x_i\}_{i=T_1}^{T_2} \subset K, \ x_{T_1} = y, \ x_{T_2} = z$$

and

$$\sum_{i=T_1}^{T_2-1} u_i(x_i, x_{i+1}) \to \max,$$

$$\{x_i\}_{i=T_1}^{T_2} \subset K, \ x_{T_1} = y.$$

The results of this section were obtained from [81].
For each integer $t \geq 0$ set

$$\|u_t\| = \sup\{|u_t(z)| : z \in \Omega\} \tag{2.30}$$

and assume that

$$\sup\{\|u_t\| : t = 0, 1, \dots\} < \infty. \tag{2.31}$$

In the sequel we assume that supremum of empty set is $-\infty$.
For each $y \in K$ and each pair of integers T_1, T_2 satisfying $T_1 < T_2$ set

$$U(y, T_1, T_2) = \sup \left\{ \sum_{t=T_1}^{T_2-1} u_t(x_t, x_{t+1}) : \{x_t\}_{t=T_1}^{T_2} \text{ is a program and } x_{T_1} = y \right\}.$$

$$\tag{2.32}$$

Let $y, \tilde{y} \in K$ and integers T_1, T_2 satisfy $T_1 < T_2$. Set

$$U(y, \tilde{y}, T_1, T_2) = \sup \left\{ \sum_{t=T_1}^{T_2-1} u_t(x_t, x_{t+1}) : \right.$$

$$\left. \{x_t\}_{t=T_1}^{T_2} \text{ is a program and } x_{T_1} = y, \ x_{T_2} = \tilde{y} \right\}. \tag{2.33}$$

Let the integers T_1, T_2 satisfy $T_1 < T_2$. Set

$$\widehat{U}(T_1, T_2) = \sup \left\{ \sum_{t=T_1}^{T_2-1} u_t(x_t, x_{t+1}) : \{x_t\}_{t=T_1}^{T_2} \text{ is a program} \right\}. \tag{2.34}$$

Upper semicontinuity of the functions u_t, $t = 0, 1, \ldots$ implies the following three results.

Proposition 2.2 *For each $z \in K$ and each pair of integers $T_1 < T_2$ such that $U(z, T_1, T_2)$ is finite there exists a program $\{x_t\}_{t=T_1}^{T_2}$ such that $x_{T_1} = z$ and*

$$\sum_{t=T_1}^{T_2-1} u_t(x_t, x_{t+1}) = U(z, T_1, T_2).$$

Proposition 2.3 *Let $y_0, \tilde{y}_0 \in K$, integers T_1, T_2 satisfy $T_1 < T_2$ and let*

$$U(y_0, \tilde{y}_0, T_1, T_2)$$

be finite. Then there exists a program $\{x_t\}_{t=T_1}^{T_2}$ such that

$$x_{T_1} = y_0, \ x_{T_2} = \tilde{y}_0,$$

$$\sum_{t=T_1}^{T_2-1} u_t(x_t, x_{t+1}) = U(y_0, \tilde{y}_0, T_1, T_2).$$

Proposition 2.4 *Let the integers T_1, T_2 satisfy $T_1 < T_2$ and let $\widehat{U}(T_1, T_2)$ be finite. Then there exists a program $\{x_t\}_{t=T_1}^{T_2}$ such that*

$$\sum_{t=T_1}^{T_2-1} u_t(x_t, x_{t+1}) = \widehat{U}(T_1, T_2).$$

In this section we suppose that the following assumption holds.

(A)   There exists a natural number $\bar{L}$ such that for each $y, z \in K$ there is a program $\{x_t\}_{t=0}^{\bar{L}}$ such that $x_0 = y$ and $x_{\bar{L}} = z$.
       Note that in [72, 73, 75] the case where $\bar{L} = 1$ was studied. In view of Proposition 2.1, assumption (A) holds for the forest management problem.

Proposition 2.5 *Let the integers T_1, T_2 satisfy $T_1 \leq T_2 - \bar{L}$ and let $y, z \in K$. Then $U(y, z, T_1, T_2)$ is finite.*

Proof If $T_2 - T_1 = \bar{L}$, then by Assumption (A), $U(y, z, T_1, T_2)$ is finite.
    Assume that $T_2 - T_1 > \bar{L}$. By (A) there exists a program $\{x_t\}_{t=T_1}^{T_2-\bar{L}}$ such that $x_{T_1} = y$ and a program $\{x_t\}_{t=T_2-\bar{L}}^{T_2}$ such that $x_{T_2} = z$. This completes the proof of Proposition 2.5.

In Sect. 2.4 we prove the following two results.

Theorem 2.6 *There exists $M > 0$ such that for each $x_0 \in K$ there exists a program $\{\bar{x}_t\}_{t=0}^{\infty}$ such that $\bar{x}_0 = x_0$ and that for each pair of integers $T_1, T_2 \geq 0$ satisfying $T_1 < T_2$,*

$$\left| \sum_{t=T_1}^{T_2-1} u_t(\bar{x}_t, \bar{x}_{t+1}) - \widehat{U}(T_1, T_2) \right| \leq M.$$

Theorem 2.7 *Assume that $\{x_t\}_{t=0}^{\infty}$ is a program and there exists $M_0 > 0$ such that for each integer $T > 0$,*

$$\sum_{t=0}^{T-1} u_t(x_t, x_{t+1}) \geq U(x(0), x(T), 0, T) - M_0.$$

Then there exists $M_1 > 0$ such that for each pair of integers $T_1 \geq 0$, $T_2 > T_1$,

$$\left| \sum_{t=T_1}^{T_2-1} u_t(x_t, x_{t+1}) - \widehat{U}(T_1, T_2) \right| \leq M_1.$$

The program $\{\bar{x}_t\}_{t=0}^{\infty}$ whose existence is guaranteed by Theorem 2.6 in infinite horizon optimal control is considered as an (approximately) optimal program [76, 84, 86].

Proposition 2.8 *Let $x_0 \in K$ and let a program $\{\bar{x}_t\}_{t=0}^{\infty}$ and $M > 0$ be as guaranteed by Theorem 2.6. Assume that $\{x_t\}_{t=0}^{\infty}$ is a program. Then either the sequence*

$$\left\{ \sum_{t=0}^{T-1} u_t(x_t, x_{t+1}) - \sum_{t=0}^{T-1} u_t(\bar{x}_t, \bar{x}_{t+1}) \right\}_{T=1}^{\infty}$$

is bounded or

$$\sum_{t=0}^{T-1} u_t(x_t, x_{t+1}) - \sum_{t=0}^{T-1} u_t(\bar{x}_t, \bar{x}_{t+1}) \to -\infty \text{ as } T \to \infty. \tag{2.35}$$

Proof Assume that the sequence

$$\left\{ \sum_{t=0}^{T-1} u_t(x_t, x_{t+1}) - \sum_{t=0}^{T-1} u_t(\bar{x}_t, \bar{x}_{t+1}) \right\}_{T=1}^{\infty}$$

is not bounded. Then by Theorem 2.6,

$$\liminf_{T\to\infty}\left[\sum_{t=0}^{T-1}u_t(x_t,x_{t+1})-\sum_{t=0}^{T-1}u_t(\bar{x}_t,\bar{x}_{t+1})\right]=-\infty.$$

Let $Q>0$. Then there exists an integer $T_0>0$ such that

$$\sum_{t=0}^{T_0-1}u_t(x_t,x_{t+1})-\sum_{t=0}^{T_0-1}u_t(\bar{x}_t,\bar{x}_{t+1})<-Q-M. \tag{2.36}$$

By (2.36), the choice of $\{\bar{x}_t\}_{t=0}^{\infty}$ and Theorem 2.6 for each integer $T>T_0$,

$$\sum_{t=0}^{T-1}u_t(x_t,x_{t+1})-\sum_{t=0}^{T-1}u_t(\bar{x}_t,\bar{x}_{t+1})=\sum_{t=0}^{T_0-1}u_t(x_t,x_{t+1})$$

$$-\sum_{t=0}^{T_0-1}u_t(\bar{x}_t,\bar{x}_{t+1})+\sum_{t=T_0}^{T-1}u_t(x_t,x_{t+1})-\sum_{t=T_0}^{T-1}u_t(\bar{x}_t,\bar{x}_{t+1})$$

$$<-Q-M+\widehat{U}(T_0,T)-\sum_{t=T_0}^{T-1}u_t(\bar{x}_t,\bar{x}_{t+1})\le-Q.$$

Since Q is any positive number we conclude that (2.35) is true. Proposition 2.8 is proved.

Now assume that $u_t=u_0$, $t=0,1,\dots$. The following result will be proved in Sect. 2.5.

Theorem 2.9 *Let $M>0$ be as guaranteed by Theorem 2.6. There exists $\mu=\lim_{p\to\infty}\widehat{U}(0,p)/p$ and*

$$|p^{-1}\widehat{U}(0,p)-\mu|\le2M/p \text{ for all natural numbers } p.$$

The following result will be proved in Sect. 2.6.

Theorem 2.10 *Let K be a compact subset of a normed space $(Z,\|\cdot\|)$, the set Ω be convex, $u_t=u_0$, $t=0,1,\dots$, the function u_0 be concave, and let μ be as guaranteed by Theorem 2.9. Then*

$$\mu=\sup\{u_0(x,x):\ x\in K \text{ such that } (x,x)\in\Omega\}.$$

In view of Proposition 2.1, the results stated in this section are applied to the forest management problem which is discussed in Sect. 2.1. These results allow us to use new optimality criterions for the forest management problem when the functions

u_t are not concave. In particular, the program $\{\bar{x}_t\}_{t=0}^{\infty}$ whose existence is guaranteed by Theorem 2.6 can be considered as an optimal solution of the corresponding forest management problem.

2.3 Auxiliary Results

Set

$$M = 2(2\bar{L} + 4) \sup\{\|u_t\| : t = 0, 1, \dots\}. \tag{2.37}$$

Lemma 2.11 *Let the integers* $T_1, T_2 \geq 0$ *satisfy* $T_2 \geq T_1 + \bar{L}$, $y, z \in K$, *and* $M_1 \geq 0$ *and let a program* $\{x_t\}_{t=T_1}^{T_2}$ *satisfy*

$$x_0 = y, \ x_T = z, \tag{2.38}$$

$$\sum_{t=T_1}^{T_2-1} u_t(x_t, x_{t+1}) \geq U(y, z, T_1, T_2) - M_1. \tag{2.39}$$

Then for each pair of integers $\tau_1, \tau_2 \geq 0$ *satisfying*

$$T_1 \leq \tau_1 < \tau_2 \leq T_2$$

the following inequality holds:

$$\sum_{t=\tau_1}^{\tau_2-1} u_t(x_t, x_{t+1}) \geq \widehat{U}(\tau_1, \tau_2) - M_1 - M. \tag{2.40}$$

Proof Let the integers $\tau_1, \tau_2 \geq 0$ satisfy

$$T_1 \leq \tau_1 < \tau_2 \leq T_2.$$

By Proposition 2.4 and assumption (A), there is a program $\{x_t'\}_{t=\tau_1}^{\tau_2}$ such that

$$\sum_{t=\tau_1}^{\tau_2-1} u_t(x_t', x_{t+1}') = \widehat{U}(\tau_1, \tau_2). \tag{2.41}$$

If $\tau_2 - \tau_1 \leq 2\bar{L} + 2$, then in view of (2.37) inequality (2.40) holds.
    Assume that

$$\tau_2 - \tau_1 > 2\bar{L} + 2. \tag{2.42}$$

By assumption (A) there exist programs $\{y_t\}_{t=\tau_1}^{\tau_1+\bar{L}}$, $\{y_t'\}_{t=\tau_2-\bar{L}}^{\tau_2}$ such that

$$y'_{\tau_2-\bar{L}} = x'_{\tau_2-\bar{L}}, \ y'_{\tau_2} = x_{\tau_2}, \ y_{\tau_1} = x_{\tau_1}, \ y_{\tau_1+\bar{L}} = x'_{\tau_1+\bar{L}}. \tag{2.43}$$

Set

$$z_t = x_t, \ t = T_1, \dots, \tau_1, \ z_t = y_t, \ t = \tau_1 + 1, \dots, \tau_1 + \bar{L},$$

$$z_t = x'_t, \ t = \tau_1 + \bar{L} + 1, \dots, \tau_2 - \bar{L},$$

$$z_t = y'_t, \ t = \tau_2 - \bar{L} + 1, \dots, \tau_2,$$

$$z_t = x_t \ \text{for all integers } t \text{ satisfying } \tau_2 < t \le T_2. \tag{2.44}$$

Clearly, $\{z_t\}_{t=T_1}^{T_2}$ is a program. By (2.37), (2.39), (2.41), (2.43), and (2.44),

$$-M_1 \le \sum_{t=T_1}^{T_2-1} u_t(x_t, x_{t+1}) - \sum_{t=T_1}^{T_2-1} u_t(z_t, z_{t+1})$$

$$= \sum_{t=\tau_1}^{\tau_2-1} u_t(x_t, x_{t+1}) - \sum_{t=\tau_1}^{\tau_2-1} u_t(z_t, z_{t+1})$$

$$\le \sum_{t=\tau_1}^{\tau_2-1} u_t(x_t, x_{t+1}) - \sum_{t=\tau_1}^{\tau_2-1} u_t(x'_t, x'_{t+1})$$

$$+ 2 \sup\{\|u_t\| : \ t = 0, 1, \dots\} 2(\bar{L} + 1)$$

$$\le \sum_{t=\tau_1}^{\tau_2-1} u_t(x_t, x_{t+1}) - \widehat{U}(\tau_1, \tau_2) + M$$

and

$$\sum_{t=\tau_1}^{\tau_2-1} u_t(x_t, x_{t+1}) \ge \widehat{U}(\tau_1, \tau_2) - M_1 - M.$$

Lemma 2.11 is proved.

Lemma 2.11 and (2.37) imply the following auxiliary result.

Lemma 2.12 *Let the integers* $T_1, T_2 \geq 0$ *satisfy* $T_2 > T_1$, $y \in K$, $M_1 \geq 0$ *and let a program* $\{x_t\}_{t=T_1}^{T_2}$ *satisfy*

$$x_0 = y,$$

$$\sum_{t=T_1}^{T_2-1} u_t(x_t, x_{t+1}) \geq U(y, T_1, T_2) - M_1.$$

Then for each pair of integers $\tau_1, \tau_2 \geq 0$ *satisfying*

$$T_1 \leq \tau_1 < \tau_2 \leq T_2,$$

$$\sum_{t=\tau_1}^{\tau_2-1} u_t(x_t, x_{t+1}) \geq \widehat{U}(\tau_1, \tau_2) - M_1 - M.$$

2.4 Proofs of Theorems 2.6 and 2.7

Proof of Theorem 2.6 Let $x_0 \in K$. By Proposition 2.2, for each natural number k there exists a program $\{x_t^{(k)}\}_{t=0}^k$ such that

$$x_0^{(k)} = x_0, \quad \sum_{t=0}^{k-1} u_t(x_t^{(k)}, x_{t+1}^{(k)}) = U(x_0, 0, k). \tag{2.45}$$

By (2.45) and Lemma 2.12 the following property holds:

(P3) For each integer $k \geq 0$ and each pair of integers $\tau_1, \tau_2 \geq 0$ satisfying $0 \leq \tau_1 < \tau_2 \leq k$,

$$\sum_{t=\tau_1}^{\tau_2-1} u_t(x_t^{(k)}, x_{t+1}^{(k)}) \geq \widehat{U}(T_1, T_2) - M.$$

Clearly, there exists a strictly increasing sequence of natural numbers $\{k_j\}_{j=1}^\infty$ such that for each integer $t \geq 0$ there exists

$$\bar{x}_t = \lim_{j \to \infty} x_t^{(k_j)}. \tag{2.46}$$

In view of (2.45) and (2.46),

$$\bar{x}_0 = x_0. \tag{2.47}$$

It follows from (P3), (2.46), and upper semicontinuity of the functions u_t, $t = 0, 1, \ldots$ that for each pair of integers $\tau_1, \tau_2 \geq 0$ satisfying $\tau_1 < \tau_2$,

$$\left| \sum_{t=\tau_1}^{\tau_2-1} u_t(\bar{x}_t, \bar{x}_{t+1}) - \widehat{U}(T_1, T_2) \right| \leq M.$$

Theorem 2.6 is proved.

Proof of Theorem 2.7 By Lemma 2.11, for all integers $\tau_1, \tau_2 \geq 0$ satisfying $0 \leq \tau_1 < \tau_2$,

$$\sum_{t=\tau_1}^{\tau_2-1} u_t(x_t, x_{t+1}) - \widehat{U}(\tau_1, \tau_2) \geq -M_0 - M.$$

This completes the proof of Theorem 2.7.

2.5 Proof of Theorem 2.9

Set $u = u_0$.

Let $x_0 \in K$ and let $\{\bar{x}_t\}_{t=0}^{\infty}$ be as guaranteed by Theorem 2.6. Then for each pair of integers $T_1, T_2 \geq 0$ satisfying $T_1 < T_2$,

$$\left| \sum_{t=T_1}^{T_2-1} u(\bar{x}_t, \bar{x}_{t+1}) - \widehat{U}(T_1, T_2) \right| \leq M. \tag{2.48}$$

Choose $\Delta > 0$ such that

$$\Delta > \|u_0\|. \tag{2.49}$$

Let p be a natural number. We show that for all sufficiently large natural numbers T,

$$\left| p^{-1}\widehat{U}(0, p) - T^{-1} \sum_{t=0}^{T-1} u(\bar{x}_t, \bar{x}_{t+1}) \right| \leq 2M/p. \tag{2.50}$$

Assume that $T \geq p$ is a natural number. Then there exist integers q, s such that

$$q \geq 1, \ 0 \leq s < p, \ T = pq + s. \tag{2.51}$$

It follows from (2.51) that

$$T^{-1} \sum_{t=0}^{T-1} u(\bar{x}_t, \bar{x}_{t+1}) - p^{-1} \widehat{U}(0, p)$$

$$= T^{-1} \left(\sum_{t=0}^{pq-1} u(\bar{x}_t, \bar{x}_{t+1}) \right.$$

$$\left. + \sum \{u(\bar{x}_t, \bar{x}_{t+1}) : t \text{ is an integer such that } pq \le t \le T-1\} \right) - p^{-1} \widehat{U}(0, p)$$

$$= T^{-1} \sum \{u(\bar{x}_t, \bar{x}_{t+1}) : t \text{ is an integer such that } pq \le t \le T-1\}$$

$$+ (T^{-1}pq)(pq)^{-1} \sum_{t=0}^{q-1} \sum_{t=ip}^{(i+1)p-1} u(\bar{x}_t, \bar{x}_{t+1}) - p^{-1} \widehat{U}(0, p)$$

$$= (T^{-1}pq)(pq)^{-1} \left[\sum_{i=0}^{q-1} \left(\sum_{t=ip}^{(i+1)p-1} u(\bar{x}_t, \bar{x}_{t+1}) - \widehat{U}(0, p) \right) + q\widehat{U}(0, p) \right]$$

$$- p^{-1} \widehat{U}(0, p)$$

$$+ T^{-1} \left\{ \sum u(\bar{x}_t, \bar{x}_{t+1}) : t \text{ is an integer such that } pq \le t \le T-1 \right\}.$$

$$(2.52)$$

By (2.48), (2.49), (2.51), and (2.52),

$$\left| T^{-1} \sum_{t=0}^{T-1} u(\bar{x}_t, \bar{x}_{t+1}) - p^{-1} \widehat{U}(0, p) \right|$$

$$\le T^{-1} p\Delta + (pq)^{-1} qM + |\widehat{U}(0, p)||q/T - 1/p|$$

$$\le T^{-1} p\Delta + M/p + |\widehat{U}(0, p)|s(pT)^{-1} \to M/p \text{ as } T \to \infty. \qquad (2.53)$$

Since p is any natural number we conclude that

$$\left\{ T^{-1} \sum_{t=0}^{T-1} u(\bar{x}_t, \bar{x}_{t+1}) \right\}_{T=1}^{\infty}$$

is a Cauchy sequence. Clearly, there exists

$$\lim_{T \to \infty} T^{-1} \sum_{t=0}^{T-1} u(\bar{x}_t, \bar{x}_{t+1}),$$

and that in view of (2.53) for each natural number p,

$$|p^{-1}\widehat{U}(0, p) - \lim_{T\to\infty} T^{-1} \sum_{t=0}^{T-1} u(\bar{x}_t, \bar{x}_{t+1})| \leq 2M/p. \tag{2.54}$$

Since (2.54) is true for any natural number p we obtain that

$$\lim_{T\to\infty} T^{-1} \sum_{t=0}^{T-1} u(\bar{x}_t, \bar{x}_{t+1}) = \lim_{p\to\infty} \widehat{U}(0, p)/p. \tag{2.55}$$

Set

$$\mu = \lim_{p\to\infty} \widehat{U}(0, p)/p. \tag{2.56}$$

By (2.54)–(2.56), for all natural numbers p,

$$|p^{-1}\widehat{U}(0, p) - \mu| \leq 2M/p.$$

Theorem 2.9 is proved.

2.6   Proof of Theorem 2.10

Put $u = u_0$. By Proposition 2.4 and assumption (A), for each integer $T > 0$ there is a program $\{x_t^{(T)}\}_{t=0}^{T}$ such that

$$\sum_{t=0}^{T-1} u(x_t^{(T)}, x_{t+1}^{(T)}) = \widehat{U}(0, T). \tag{2.57}$$

Set

$$x^{(T)} = T^{-1} \sum_{t=0}^{T-1} x_t^{(T)}, \quad y^{(T)} = T^{-1} \sum_{t=0}^{T-1} x_{t+1}^{(T)}. \tag{2.58}$$

Clearly,

$$(x^{(T)}, y^{(T)}) \in \Omega, \quad Tu(x^{(T)}, y^{(T)}) \geq \widehat{U}(0, T),$$

$$u(x^{(T)}, y^{(T)}) \geq T^{-1}\widehat{U}(0, T),$$

$$\|x^{(T)} - y^{(T)}\| \leq 2T^{-1} \sup\{\|z\| : z \in K\}.$$

There is a strictly increasing sequence of natural numbers $\{T_k\}_{k=1}^{\infty}$ such that there exist

$$\lim_{k \to \infty} x^{(T_k)} = \bar{x}, \quad \lim_{k \to \infty} y^{(T_k)} = \bar{y}.$$

Clearly,

$$\bar{x} = \bar{y}, \quad (\bar{x}, \bar{x}) \in \Omega,$$

$$u(\bar{x}, \bar{x}) \geq \lim_{T \to \infty} T^{-1}\widehat{U}(0, T) = \mu.$$

On the other hand, it is easy to see that

$$u(z, z) \leq \mu$$

for all $z \in K$ such that $(z, z) \in \Omega$. Theorem 2.10 is proved.

2.7 Infinite Horizon Problems with Discounting

We consider the optimal control problem of Sect. 2.2 and continue to use the notation, definitions, and assumptions introduced there.

In this section we suppose that the following equality holds:

$$\lim_{t \to \infty} \sup\{|u_t(z)| : z \in \Omega\} = 0. \tag{2.59}$$

By (A) and Proposition 2.2, for each $y \in K$ and each natural number T there is a program $\{x_t^{(y,T)}\}_{t=0}^{T}$ such that

$$x_0^{(y,T)} = y, \tag{2.60}$$

$$\sum_{t=0}^{T-1} u_t(x_t^{(y,T)}, x_{t+1}^{(y,T)}) = U(y, 0, T). \tag{2.61}$$

In Sect. 2.9 we prove the following result.

Theorem 2.13 *For any $y \in K$ there exists a program $\{x_t^{(y)}\}_{t=0}^{\infty}$ such that $x_0^{(y)} = y$ and the following property holds:*

For each $\epsilon > 0$ there exists a natural number τ such that for each $y \in K$ and each integer $T \geq \tau$,

$$\left| \sum_{t=0}^{T-1} u_t(x_t^{(y)}, x_{t+1}^{(y)}) - U(y, 0, T) \right| \le \epsilon.$$

The next corollary easily follows from Theorem 2.13.

Corollary 2.14 *Let $y \in K$. Then for any program $\{x_t\}_{t=0}^{\infty}$ satisfying $x_0 = y$,*

$$\limsup_{T \to \infty} \sum_{t=0}^{T-1} \left[u_t(x_t, x_{t+1}) - u_t(x_t^{(y)}, x_{t+1}^{(y)}) \right] \le 0.$$

Note that the program $\{x_t\}_{t=0}^{\infty}$ which exists by Corollary 2.14 is called in the literature as an overtaking optimal program [76, 84, 86].

Example 2.15 Let $w : \Omega \to [0, \infty)$ be a bounded upper semicontinuous function, $\{\rho_t\}_{t=0}^{\infty} \subset (0, 1)$ satisfy

$$\lim_{t \to \infty} \rho_t = 0, \tag{2.62}$$

and let $u_t = \rho_t w$, $t = 0, 1, \ldots$. Then (2.59) holds. In the literature it is also considered an optimality criterion with $\rho_t = \alpha^t$, $t = 0, 1, \ldots$ where $\alpha \in (0, 1)$. In this case for any program $\{x_t\}_{t=0}^{\infty}$, $\sum_{t=0}^{\infty} \alpha^t w(x_t, x_{t+1}) < \infty$. This convergence does not hold in the general case with $\{\rho_t\}_{t=0}^{\infty} \subset (0, 1)$ satisfying (2.62). Therefore in the general case the existence problem of an overtaking optimal program is more difficult and less understood.

The results of this section were obtained from [80].

2.8 An Auxiliary Result for Theorem 2.13

Recall that for any integer $t \ge 0$,

$$\|u_t\| = \sup\{|u_t(z)| : z \in \Omega\}. \tag{2.63}$$

Lemma 2.16 *Let $\epsilon > 0$. Then there exists a natural number τ such that for each $y \in K$ and each pair of integers $T_1 \ge \tau$ and $T_2 \ge T_1 + \bar{L}$,*

$$\sum_{t=0}^{T_1-1} u_t \left(x_t^{(y, T_2)}, x_{t+1}^{(y, T_2)} \right) \ge U(y, 0, T_1) - \epsilon.$$

Proof By (2.59) and (2.63), there exists a natural number τ such that

$$\|u_t\| \le \epsilon(4\bar{L})^{-1} \text{ for all integers } t \ge \tau. \tag{2.64}$$

Assume that $y \in K$ and that integers

$$T_1 \geq \tau, \ T_2 \geq T_1 + \bar{L}. \tag{2.65}$$

In view of (A), there exists a program $\{x_t\}_{t=T_1}^{T_1+\bar{L}}$ such that

$$x_{T_1} = x_{T_1}^{(y,T_1)}, \ x_{T_1+\bar{L}} = x_{T_1+\bar{L}}^{(y,T_2)}. \tag{2.66}$$

Set

$$x_t = x_t^{(y,T_1)}, \ t = 0, \ldots, T_1 - 1, \tag{2.67}$$

$$x_t = x_t^{(y,T_2)} \text{ for all integers } t \text{ satisfying } T_1 + \bar{L} < t \leq T_2.$$

Clearly, $\{x_t\}_{t=0}^{T_2}$ is a program and

$$x_0 = y. \tag{2.68}$$

By (2.60), (2.61), (2.63)–(2.66), and (2.68),

$$0 \leq \sum_{t=0}^{T_2-1} u_t(x_t^{(y,T_2)}, x_{t+1}^{(y,T_2)}) - \sum_{t=0}^{T_2-1} u_t(x_t, x_{t+1})$$

$$= \sum_{t=0}^{T_1+\bar{L}-1} u_t(x_t^{(y,T_2)}, x_{t+1}^{(y,T_2)}) - \sum_{t=0}^{T_1+\bar{L}-1} u_t(x_t, x_{t+1})$$

$$\leq \sum_{t=0}^{T_1-1} u_t(x_t^{(y,T_2)}, x_{t+1}^{(y,T_2)}) - \sum_{t=0}^{T_1-1} u_t(x_t^{(y,T_1)}, x_{t+1}^{(y,T_1)}) + 2 \sum_{t=T_1}^{T_1+\bar{L}-1} \|u_t\|$$

$$\leq \sum_{t=0}^{T_1-1} u_t(x_t^{(y,T_2)}, x_{t+1}^{(y,T_2)}) - U(y, 0, T_1) + 2\bar{L}(\epsilon(4\bar{L})^{-1})$$

and

$$\sum_{t=0}^{T_1-1} u_t(x_t^{(y,T_2)}, x_{t+1}^{(y,T_2)}) \geq U(y, 0, T_1) - \epsilon.$$

Lemma 2.16 is proved.

2.9   Proof of Theorem 2.13

Let $y \in K$. Using the diagonalization process and the compactness of K we obtain a strictly increasing sequence of natural numbers $\{T_k\}_{k=1}^{\infty}$ such that for any integer $t \geq 0$ there exists

$$x_t^{(y)} = \lim_{k \to \infty} x_t^{(y,T_k)}. \tag{2.69}$$

Clearly, $\{x_t^y\}_{t=0}^{\infty}$ is a program for all $y \in K$.

Let $\epsilon > 0$ and let a natural number τ be as guaranteed by Lemma 2.16. Assume that an integer $T \geq \tau$ and $y \in K$. Then for all sufficiently large natural numbers k,

$$\sum_{t=0}^{T-1} u_t(x_t^{(y,T_k)}, x_{t+1}^{(y,T_k)}) \geq U(y, 0, T) - \epsilon.$$

By the inequality above, (2.69) and upper semicontinuity of the functions u_t, $t = 0, 1, \ldots,$

$$\sum_{t=0}^{T-1} u_t(x_t^{(y)}, x_{t+1}^{(y)}) \geq U(y, 0, T) - \epsilon.$$

Theorem 2.13 is proved.

2.10   An Application to the Forest Management Problem

We consider the forest management problem discussed in Sect. 2.1 using the notation, definitions, and assumptions introduced there.

Recall that for each $(x, y) \in \Omega$,

$$V(x, y) = (v_1(x, y), \ldots, v_k(x, y)),$$

where for $i = 1, \ldots, k$,

$$v_i(x, y) = x_i^{n_i+1} + x_i^{n_i} - y_i^{n_i+1},$$

and that

$$\Delta_0 = \left\{ u \in R_+^k : \sum_{i=1}^{k} u_i \leq S \right\}.$$

In this section we assume that a benefit at moment $t = 0, 1, \ldots$ is represented by an upper semicontinuous function $w_t : \Delta_0 \to R^1$ and at a moment $t = 0, 1, \ldots,$ $w_t(V(x, y))$ is the benefit obtained today if the forest today is x and the forest tomorrow is y, where $(x, y) \in \Omega$. In view of Proposition 2.1, (A) holds.

Assume that

$$\sup\{|w_t(z)| : z \in \Delta_0, \ t = 0, 1, \ldots\} < \infty.$$

It is easy now to see that all the results of Sect. 2.2 hold for our model with $u_t = w_t \circ V, t = 0, 1, \ldots.$

If

$$\sup\{|w_t(z)| : z \in \Delta_0\} \to 0 \text{ as } t \to \infty,$$

then all the results of Sect. 2.7 hold for our model with $u_t = w_t \circ V, t = 0, 1, \ldots.$

Chapter 3
Turnpike Properties

In this chapter we consider a discrete-time optimal control problem. The forest management problem is its particular case. We establish turnpike results for approximate solutions and the stability of the turnpike phenomenon. We prove the existence of solutions of the corresponding infinite horizon problems and show the equivalence of different optimality criterions for these problems. Most results of this chapter were obtained from [83].

3.1 Preliminaries and Main Results

We continued to study the class of optimal control problems introduced in Sect. 2.2 using the same notation, definitions, and assumptions.

Let (K, ρ) be a compact metric space and let Ω be a nonempty closed subset of $K \times K$.

A sequence $\{x_t\}_{t=0}^{\infty} \subset K$ is called an (Ω)-program (or a program if Ω is understood) if $(x_t, x_{t+1}) \in \Omega$ for all integers $t \geq 0$.

Let the integers T_1, T_2 satisfy $T_1 < T_2$. A sequence $\{x_t\}_{t=T_1}^{T_2} \subset K$ is called an (Ω)-program (or a program if Ω is understood) if $(x_t, x_{t+1}) \in \Omega$ for all integers t satisfying $T_1 \leq t < T_2$.

For each integer $t \geq 0$, let $u_t : \Omega \to R^1$ be a bounded upper semicontinuous function.

For each pair of integers T_1, T_2 satisfying $0 \leq T_1 < T_2$ and each $y, z \in K$ we consider the optimization problems

$$\sum_{i=T_1}^{T_2-1} u_i(x_i, x_{i+1}) \to \max, \qquad (P_1)$$

$$\{x_i\}_{i=T_1}^{T_2} \subset K \text{ is a program, } x_{T_1} = y, \; x_{T_2} = z,$$

$$\sum_{i=T_1}^{T_2-1} u_i(x_i, x_{i+1}) \to \max, \tag{P_2}$$

$$\{x_i\}_{i=T_1}^{T_2} \subset K \text{ is a program, } x_{T_1} = y$$

and

$$\sum_{i=T_1}^{T_2-1} u_i(x_i, x_{i+1}) \to \max, \tag{P_3}$$

$$\{x_i\}_{i=T_1}^{T_2} \subset K \text{ is a program.}$$

The interest in discrete-time optimal problems of these types stems from the study of various optimization problems which can be reduced to it, e.g., continuous-time control systems which are represented by ordinary differential equations whose cost integrand contains a discounting factor [16], tracking problems in engineering [44, 87], the study of Frenkel-Kontorova model related to dislocations in one-dimensional crystals [6, 71], the analysis of a long slender bar of a polymeric material under tension in [19, 45, 50, 51], and models of economic growth [84, 86]. See also [72, 73] where these problems were studied with $\Omega = K \times K$.

For each integer $t \geq 0$ set

$$\|u_t\| = \sup\{|u_t(z)| : \; z \in \Omega\} \tag{3.1}$$

and assume that

$$\sup\{\|u_t\| : \; t = 0, 1, \dots\} < \infty. \tag{3.2}$$

In the sequel we assume that supremum over empty set is $-\infty$.

For every point $y \in K$ and every pair of integers T_1, T_2 satisfying $T_1 < T_2$ set

$$U(y, T_1, T_2) = \sup \left\{ \sum_{t=T_1}^{T_2-1} u_t(x_t, x_{t+1}) : \; \{x_t\}_{t=T_1}^{T_2} \text{ is a program and } x_{T_1} = y \right\}. \tag{3.3}$$

Let a point $y, \tilde{y} \in K$ be given and let the integers T_1, T_2 satisfy $T_1 < T_2$. Set

$$U(y, \tilde{y}, T_1, T_2) = \sup \left\{ \sum_{t=T_1}^{T_2-1} u_t(x_t, x_{t+1}) : \right.$$

$$\left. \{x_t\}_{t=T_1}^{T_2} \text{ is a program and } x_{T_1} = y, \ x_{T_2} = \tilde{y} \right\}. \tag{3.4}$$

Let the integers T_1, T_2 satisfy $T_1 < T_2$. Define

$$\hat{U}(T_1, T_2) = \sup \left\{ \sum_{t=T_1}^{T_2-1} u_t(x_t, x_{t+1}) : \ \{x_t\}_{t=T_1}^{T_2} \text{ is a program} \right\}. \tag{3.5}$$

In this chapter we suppose that the following assumption, introduced in Sect. 2.2, holds.

(A) There exists a natural number $\bar{L}$ such that for every pair of points $y, z \in K$ there exists a program $\{x_t\}_{t=0}^{\bar{L}}$ which satisfies $x_0 = y$ and $x_{\bar{L}} = z$.

This assumption holds for our forest management problems. Note that all the results stated in Sect. 2.2 (Theorems 2.6, 2.7, 2.9, and 2.10 and Propositions 2.5 and 2.8) are valid.

We suppose that $u_t = u_0$ for all integers $t \geq 0$. Set

$$u = u_0 \text{ and } \|u\| = \sup\{|u(z)| : z \in \Omega\}. \tag{3.6}$$

Let us consider the constant μ defined by Theorem 2.9 and suppose that

$$\mu = \mu(u) = \sup\{u(x, x) : x \in K \text{ and } (x, x) \in \Omega\}. \tag{3.7}$$

By upper semicontinuity of u, there exists a point $\bar{x} \in K$ such that

$$(\bar{x}, \bar{x}) \in \Omega \text{ and } u(\bar{x}, \bar{x}) = \mu. \tag{3.8}$$

Denote by $\mathrm{Card}(A)$ the cardinality of a set A.

A program $\{x_t\}_{t=0}^{\infty}$ is called (u, Ω)-good $((u)$-good if the set Ω is understood or good if the pair (u, Ω) is understood) if the sequence

$$\left\{ \sum_{t=0}^{T-1} \{u(x_t, x_{t+1}) - \mu\} \right\}_{T=1}^{\infty}$$

is bounded.

A program $\{x_t\}_{t=0}^{\infty}$ is called (u, Ω)-bad $((u)$-bad if the set Ω is understood or bad if the pair (u, Ω) is understood) if $\sum_{t=0}^{T-1}(u(x_t, x_{t+1}) - \mu) \to -\infty$ as $T \to \infty$.

Theorem 2.9, Proposition 2.8, and Theorem 2.6 imply the following result.

Proposition 3.1 *Let $\{x_t\}_{t=0}^{\infty}$ be a program. Then it is either good or bad. For every point $x_0 \in K$ there exists a good program $\{x_t\}_{t=0}^{\infty}$.*

We study the structure of approximate solutions of problems (P_1)-(P_3) and suppose that the following assumptions hold.

(B1) For every good program $\{x_t\}_{t=0}^{\infty}$, $\lim_{t\to\infty} x_t = \bar{x}$.
(B2) The function u is continuous at the point $(\bar{x}, \bar{x})$ with respect to Ω.
(B3) There exists a natural number L_* such that for every positive number ϵ there exists a positive number δ such that for every pair of points $y_1, y_2 \in K$ which satisfies

$$\rho(y_1, \bar{x}), \ \rho(\bar{x}, y_2) \le \delta$$

there exists a program $\{x_t\}_{t=0}^{L_*}$ such that $x_0 = y_1$, $x_{L_*} = y_2$ and $\rho(x_t, \bar{x}) \le \epsilon$, $t = 0, \ldots, L_*$.

We will prove the following results which describe the structure of approximate solutions of problems (P_1)–(P_3).

Theorem 3.2 *Let ϵ be a positive number. Then there exist an integer $\tau > 0$ and a positive number δ such that for every natural number $T > 2\tau$ and every program $\{x_t\}_{t=0}^{T}$ which satisfy*

$$\sum_{t=0}^{T-1} u(x_t, x_{t+1}) \ge U(x_0, x_T, 0, T) - \delta$$

there exist integers $\tau_1 \in [0, \tau]$, $\tau_2 \in [T - \tau, T]$ such that

$$\rho(x_t, \bar{x}) \le \epsilon, \ t = \tau_1, \ldots, \tau_2.$$

Moreover $\tau_1 = 0$, if $\rho(x_0, \bar{x}) \le \delta$ and $\tau_2 = T$ if $\rho(x, \bar{x}) \le \delta$.

Theorem 3.2 is proved in Sect. 3.3.

Theorem 3.3 *Let ϵ, M be positive numbers. Then there exist integers τ, $M_1 > 0$ such that for every natural number $T > \tau$ and every program $\{x_t\}_{t=0}^{T}$ which satisfies*

$$\sum_{t=0}^{T-1} u(x_t, x_{t+1}) \ge U(x_0, x_T, 0, T) - M$$

the inequality

$$Card(\{t \in \{0, \ldots, T\} : \ \rho(x_t, \bar{x}) > \epsilon\}) < M_1$$

holds.

Theorem 3.3 is proved in Sect. 3.4.

Theorem 3.4 *Let $z \in K$. Then there exists a program $\{x_t\}_{t=0}^{\infty}$ such that $x_0 = z$ and that for each program $\{y_t\}_{t=0}^{\infty}$ satisfying $y_0 = z$,*

$$\limsup_{T \to \infty} \sum_{t=0}^{T-1} [u(y_t, y_{t+1}) - u(x_t, x_{t+1})] \leq 0. \tag{3.9}$$

Theorem 3.4 is proved in Sect. 3.5.

A program $\{x_t\}_{t=0}^{\infty}$ is called (u, Ω)-overtaking optimal $((u)$-overtaking optimal if the set Ω is understood or overtaking optimal if the pair (u, Ω) is understood) if

$$\limsup_{T \to \infty} \sum_{t=0}^{T-1} [u(y_t, y_{t+1}) - u(x_t, x_{t+1})] \leq 0$$

for all programs $\{y_t\}_{t=0}^{\infty}$ satisfying $x_0 = y_0$ [76, 84, 86].

In other words Theorem 3.4 establishes the existence of an overtaking optimal program for any initial state.

We also use another optimality criterion which was introduced and applied in [6, 84, 86].

A program $\{x_t\}_{t=0}^{\infty}$ is (u, Ω)-locally optimal $((u)$-locally optimal if the set Ω is understood or locally optimal if the pair (u, Ω) is understood) if for any integer $T > 0$

$$\sum_{t=0}^{T-1} u(x_t, x_{t+1}) = U(x_0, x_T, 0, T).$$

The following result will be proved in Sect. 3.6.

Theorem 3.5 *A program $\{x_t\}_{t=0}^{\infty}$ is locally optimal if and only if it is overtaking optimal.*

The next result is a generalization of Theorem 3.2 which will be proved in Sect. 3.7.

Theorem 3.6 *Let ϵ be a positive number. Then there exist an integer $\tau > 0$ and a positive number δ such that for every natural number $T > 2\tau$ and every program $\{x_t\}_{t=0}^{T}$ satisfying*

$$\sum_{t=S}^{S+\tau-1} u(x_t, x_{t+1}) \geq U(x(S), x(S + \tau), S, S + \tau) - \delta$$

for all integers $S \in \{0, \ldots, T - \tau\}$ there exist integers τ_1, τ_2 such that

$$\tau_1 \in [0, \tau], \quad \tau_2 \in [T - \tau, T],$$

$$\rho(x_t, \bar{x}) \leq \epsilon, \ t = \tau_1, \ldots \tau_2.$$

Moreover, $\tau_1 = 0$ if $\rho(x_0, \bar{x}) \leq \delta$ and $\tau_2 = T$ if $\rho(x_T, \bar{x}) \leq \delta$.

The results of this section were obtained from [83].

3.2 Auxiliary Results

Proposition 3.7 *Let $x, y \in K$ and an integer $T > 2\bar{L}$. Then*

$$U(x, y, 0, T) \geq T\mu - 4(\bar{L} + 1)\|u\|.$$

Proof In view of (A) and (3.8), there exists a program $\{x_t\}_{t=0}^{T}$ such that

$$x_0 = x, \ x_t = \bar{x}, \ t = \bar{L}, \ldots, T - \bar{L}, \ x_T = y.$$

Combined with (3.6)–(3.8) this implies that

$$U(x, y, 0, T) \geq \sum_{t=0}^{T-1} u(x_t, x_{t+1}) \geq T\mu - 4(\bar{L} + 1)\|u\|.$$

This completes the proof of Proposition 3.7.

Lemma 3.8 *Let ϵ, M be positive numbers. Then there exists an integer $T > 2\bar{L}$ such that for every program $\{x_t\}_{t=0}^{T}$ which satisfies*

$$\sum_{t=0}^{T-1} u(x_t, x_{t+1}) \geq U(x_0, x_T, 0, T) - M,$$

there exists an integer $j \in \{1, \ldots, T\}$ for which $\rho(x_j, \bar{x}) \leq \epsilon$.

Proof Assume the contrary. Then there exist a strictly increasing sequence of natural numbers $\{T_k\}_{k=1}^{\infty}$ and a sequence of programs $\{x_t^{(k)}\}_{t=0}^{T_k}$, $k = 1, 2, \ldots$ such that for every natural number k,

$$\sum_{t=0}^{T_k-1} u\left(x_t^{(k)}, x_{t+1}^{(k)}\right) \geq U\left(x_0^{(k)}, x_{T_k}^{(k)}, 0, T_k\right) - M, \tag{3.10}$$

$$\rho\left(x_j^{(k)}, \bar{x}\right) > \epsilon, \ j = 1, \ldots, T_k. \tag{3.11}$$

In view of (3.10), for every natural number k and every pair of integers S_1, S_2 which satisfy

$$0 \leq S_1 < S_2 \leq T_k$$

we have

$$\sum_{t=S_1}^{S_2-1} u(x_t^{(k)}, x_{t+1}^{(k)}) \geq U(x_{S_1}^{(k)}, x_{S_2}^{(k)}, 0, S_2 - S_1) - M. \tag{3.12}$$

Extracting a subsequence and re-indexing if necessary we may assume that for every nonnegative integer $t \geq 0$ there is

$$x_t = \lim_{k \to \infty} x_t^{(k)}. \tag{3.13}$$

It follows from Proposition 3.7 and (3.12) that the following property holds:

(P1)   For every natural number k and every pair of integers S_1, S_2 which satisfy $0 \leq S_1$ and $S_1 + 2\bar{L} < S_2 \leq T_k$, we have

$$\sum_{t=S_1}^{S_2-1} u(x_t^{(k)}, x_{t+1}^{(k)}) \geq (S_2 - S_1)\mu - 4(\bar{L}+1)\|u\| - M.$$

Property (P1) and (3.13) imply that for every pair of integers $S_1, S_2 \geq 0$ which satisfy $S_2 - S_1 > 2\bar{L}$ we have

$$\sum_{t=S_1}^{S_2-1} u(x_t, x_{t+1}) \geq (S_2 - S_1)\mu - 4\left(\bar{L}+1\right)\|u\| - M. \tag{3.14}$$

Proposition 3.1, (3.14), and property (B1) imply that the program $\{x_t\}_{t=0}^{\infty}$ is good and that

$$\lim_{t \to \infty} \rho(x_t, \bar{x}) = 0.$$

Thus there exists a natural number S_0 such that

$$\rho(x_t, \bar{x}) < \epsilon/4 \text{ for all integers } t \geq S_0. \tag{3.15}$$

In view of (3.13), there exists a natural number k such that $T_k > S_0 + 4$ and that

$$\rho(x_{S_0}^{(k)}, x_{S_0}) \leq \epsilon/4.$$

Combined with (3.15) this implies that $\rho(x_{S_0}^{(k)}, \bar{x}) \leq \epsilon/2$. This contradicts (3.11). The contradiction we have reached proves Lemma 3.8.

Lemma 3.9 *Let ϵ be a positive number. Then there exits a positive number δ such that for every natural number $T > 2L_*$ and every pair of points $y, z \in K$ which satisfies*

$$\rho(y, \bar{x}), \; \rho(z, \bar{x}) \leq \delta, \tag{3.16}$$

the inequality

$$U(y, z, 0, T) \geq T\mu - \epsilon$$

holds.

Proof Since the function u is continuous at the point $(\bar{x}, \bar{x})$ (see assumption (B2)) there exists a number $\epsilon_0 \in (0, \epsilon)$ such that for every point $(y, z) \in \Omega$ which satisfies

$$\rho(y, \bar{x}), \ \rho(z, \bar{x}) \leq \epsilon_0$$

the inequality

$$|u(y, z) - u(\bar{x}, \bar{x})| \leq \epsilon(4L_*)^{-1} \tag{3.17}$$

holds. In view of assumption (B3), there exists a number $\delta \in (0, \epsilon_0)$ such that for every pair of points $y, z \in K$ which satisfies

$$\rho(y, \bar{x}), \ \rho(z, \bar{x}) \leq \delta$$

there exists a program $\{\xi_t\}_{t=0}^{L_*}$ for which

$$\xi_0 = y, \ \xi_{L_*} = z, \ \rho(\xi_t, \bar{x}) \leq \epsilon_0, \ t = 0, \ldots, L_*. \tag{3.18}$$

Assume that a natural number $T > 2L_*$ and that a pair of points $y, z \in K$ satisfies (3.16). It follows from (3.16) and the choice of the number δ (see (3.18)) that there exists a program $\{x_t\}_{t=0}^{T}$ such that

$$x_0 = y, \ \rho(x_t, \bar{x}) \leq \epsilon_0, \ t = 0, \ldots, L_*,$$

$$x_t = \bar{x}, \ t = L_*, \ldots, T - L_*,$$

$$x_T = z, \ \rho(x_t, \bar{x}) \leq \epsilon_0, \ t = T - L_*, \ldots, T. \tag{3.19}$$

In view of (3.19), (3.8), and the choice of ϵ_0 (see (3.17)),

$$U(y, z, 0, T) \geq \sum_{t=0}^{T-1} u(x_t, x_{t+1}) = T\mu + \sum_{t=0}^{L_*-1} [u(x_t, x_{t+1}) - u(\bar{x}, \bar{x})]$$

$$+ \sum_{t=T-L_*}^{T-1} [u(x_t, x_{t+1}) - u(\bar{x}, \bar{x})] \geq T\mu$$

$$- 2L_* \epsilon(4L_*)^{-1} \geq T\mu - \epsilon/2.$$

This completes the proof of Lemma 3.9.

Lemma 3.10 *Let ϵ be a positive number. Then there exists a number $\delta \in (0, \epsilon)$ such that for every natural number $T > 2L_*$ and every program $\{x_t\}_{t=0}^{T}$ which satisfies*

$$\rho(x_0, \bar{x}), \ \rho(x_T, \bar{x}) \leq \delta, \ \sum_{t=0}^{T-1} u(x_t, x_{t+1}) \geq U(x_0, x_T, 0, T) - \delta,$$

the inequality

$$\rho(x_t, \bar{x}) \le \epsilon, \ t = 0, 1, \ldots, T$$

holds.

Proof Let k be a natural number. In view of assumption (B2), there exists a number

$$\epsilon_k \in (0, 2^{-k}) \tag{3.20}$$

such that for every point $(y, z) \in \Omega$ which satisfies

$$\rho(y, \bar{x}), \ \rho(z, \bar{x}) \le \epsilon_k$$

the inequality

$$|u(y, z) - u(\bar{x}, \bar{x})| \le 2^{-k} \tag{3.21}$$

holds.

Lemma 3.9 implies that there exists a number $\delta_{k,1} \in (0, \epsilon_k)$ such that the following property holds:

(P2) For every natural number $T > 2L_*$ and every pair of points $y, z \in K$ which satisfies

$$\rho(y, \bar{x}), \ \rho(z, \bar{x}) \le \delta_{k,1},$$

we have

$$U(y, z, 0, T) \ge T\mu - 2^{-k}.$$

In view of assumption (B3), there exists a number $\delta_{k,2} \in (0, \epsilon_k)$ such that the following property holds:

(P3) For every pair of points $y, z \in K$ which satisfy

$$\rho(y, \bar{x}), \ \rho(z, \bar{x}) \le \delta_{k,2},$$

there exists a program $\{x_t\}_{t=0}^{L_*}$ which satisfies

$$x_0 = y, \ x_{L_*} = z$$

and

$$\rho(x_t, \bar{x}) \le \epsilon_k, \ t = 0, \ldots, L_*.$$

Choose a positive number

$$\delta_k < \min\{\delta_{k,1}\delta_{k,2}\}. \tag{3.22}$$

We may assume without loss of generality that the sequence $\{\delta_i\}_{i=1}^{\infty}$ is monotone decreasing.

Assume that the assertion of the lemma does not hold. Then for every integer $k \geq 1$ there exist an integer $T_k > 2L_*$ and a program $\{x_t^{(k)}\}_{t=0}^{T_k}$ which satisfies

$$\rho(x_0^{(k)}, \bar{x}), \ \rho(x_{T_k}^{(k)}, \bar{x}) \leq \delta_k, \tag{3.23}$$

$$\sum_{t=0}^{T_k-1} u(x_t^{(k)}, x_{t+1}^{(k)}) \geq U(x_0^{(k)}, x_{T_k}^{(k)}, 0, T_k) - \delta_k, \tag{3.24}$$

$$\max\{\rho(x_t^{(k)}, \bar{x}) : \ t = 0, \ldots, T_k\} > \epsilon. \tag{3.25}$$

Set

$$S_0 = 0, \ S_k = \sum_{i=1}^{k}(T_i + L_*) \tag{3.26}$$

for all natural numbers k. By induction define a program $\{x_t^*\}_{t=0}^{\infty}$. Set

$$x_t^* = x_t^{(1)}, \ t = 0, 1, \ldots, T_1. \tag{3.27}$$

It follows from (3.22), (3.23), and property (P3) that there exists a program $\{x_t^*\}_{t=T_1}^{T_1+L_*}$ such that

$$x_{T_1+L_*}^* = x_0^{(2)}, \ \rho(x_t^*, \bar{x}) \leq \epsilon_1, \ t = T_1, \ldots, T_1 + L_*. \tag{3.28}$$

It is clear that $\{x_t^*\}_{t=0}^{S_1}$ is a program.

Assume that $k \geq 1$ is an integer and that we have defined a program $\{x_t^*\}_{t=0}^{S_k}$ such that

$$x_{S_k}^* = x_0^{(k+1)} \tag{3.29}$$

and that for each integer $i \in \{0, \ldots, k-1\}$

$$x_{S_i+t}^* = x_t^{(i+1)}, \ t = 0, \ldots, T_{i+1},$$

$$\rho(x_{S_i+T_{i+1}+t}^*, \bar{x}) \leq \epsilon_{i+1}, \ t = 0, \ldots, L_*. \tag{3.30}$$

(It is clear that this assumption holds for $k = 1$.)

Set

$$x^*_{S_k+t} = x_t^{(k+1)}, \ t = 1, \ldots, T_{k+1}. \tag{3.31}$$

In view of (3.31), (3.23), property (P3), (3.26), and (3.22), there exists a program $\{x^*_t\}_{t=S_k+T_{k+1}}^{S_{k+1}}$ such that

$$x^*_{S_{k+1}} = x_0^{(k+2)}, \ \rho(x^*_t, \bar{x}) \le \epsilon_{k+1}, \ t = S_k + T_{k+1}, \ldots, S_{k+1}. \tag{3.32}$$

It is not difficult to see that $\{x^*_t\}_{t=0}^{S_{k+1}}$ is a program and the assumption we made for k also holds for $k + 1$. Thus the program $\{x^*_t\}_{t=0}^{\infty}$ was defined by induction such that (3.30) holds for all integers $i \ge 0$.

For every nonnegative integer k it follows from (3.30), (3.26), and (3.24), the choice of ϵ_{k+1} (see (3.21)), property (P2), (3.23), (3.20), and (3.22) that

$$\sum_{t=S_k}^{S_{k+1}-1} [u(x^*_t, x^*_{t+1}) - u(\bar{x}, \bar{x})] = \sum_{t=0}^{T_{k+1}-1} \left[u(x_t^{(k+1)}, x_{t+1}^{(k+1)}) - u(\bar{x}, \bar{x}) \right]$$

$$+ \sum_{t=S_k+T_{k+1}}^{S_{k+1}-1} [u(x^*_t, x^*_{t+1}) - u(\bar{x}, \bar{x})]$$

$$\ge U\left(x_0^{(k+1)}, x_{T_{k+1}}^{(k+1)}, 0, T_{k+1} \right) - \delta_{k+1}$$

$$- T_{k+1}\mu - 2^{-k+1}L_* \ge -2^{-k+1}(L_* + 2).$$

This implies that for every nonnegative integer k we have

$$\sum_{t=0}^{S_k+1} [u(x^*_t, x^*_{t+1}) - \mu] \ge -(L_* + 2) \sum_{i=0}^{k+1} 2^{-i+1}$$

and by Proposition 3.1 and assumption (B1), the program $\{x^*_t\}_{t=0}^{\infty}$ is good and satisfies

$$\lim_{t \to \infty} x^* = \bar{x}.$$

Combined with (3.30) this contradicts (3.25). The contradiction we have reached proves Lemma 3.10.

3.3 Proof of Theorem 3.2

We may assume that $\epsilon < 1/2$. Lemma 3.10 implies that there exists a number $\delta \in (0, \epsilon)$ such that the following property holds:

(P4)   For every natural number $T > 2L_*$ and every program $\{x_t\}_{t=0}^T$ satisfying

$$\rho(x_0, \bar{x}), \ \rho(x_T, \bar{x}) \leq \delta$$

and

$$\sum_{t=0}^{T-1} u(x_t, x_{t+1}) \geq U(x_0, x_T, 0, T) - \delta$$

the inequality $\rho(x_t, \bar{x}) \leq \epsilon$ is true for all $t = 0, \ldots, T$.

Lemma 3.8 implies that there exists an integer $T_* > 2\bar{L}$ such that the following property holds:

(P5)   For every program $\{x_t\}_{t=0}^{T_*}$ which satisfies

$$\sum_{t=0}^{T_*-1} u(x_t, x_{t+1}) \geq U(x_0, x_{T_*}, 0, T_*) - 4,$$

there exists an integer $j \in \{1, \ldots, T_*\}$ for which $\rho(x_j, \bar{x}) \leq \delta$.

Set

$$\tau = T_* + L_*. \tag{3.33}$$

Assume that a natural number $T > 2\tau$ and that a program $\{x_t\}_{t=0}^T$ satisfies

$$\sum_{t=0}^{T-1} u(x_t, x_{t+1}) > U(x_0, x_T, 0, T) - \delta. \tag{3.34}$$

In view of (3.34),

$$\sum_{t=0}^{T_*-1} u(x_t, x_{t+1}) \geq U(x_0, x_{T_*}, 0, T_*) - \delta, \tag{3.35}$$

$$\sum_{t=T-T_*}^{T-1} u(x_t, x_{t+1}) \geq U(x_{T-T_*}, x_T, 0, T_*) - \delta. \tag{3.36}$$

Property (P5), (3.35), and (3.36) imply that there exist integers $\tau_1 \in [0, T_*]$, $\tau_2 \in [T - T_*, T]$ such that

$$\rho(x_{\tau_1}, \bar{x}) \leq \delta, \quad \rho(x_{\tau_2}, \bar{x}) \leq \delta. \tag{3.37}$$

It is clear that if $\rho(x_0, \bar{x}) \leq \delta$, then we may set $\tau_1 = 0$, and if $\rho(x_T, \bar{x}) \leq \delta$, then we may set $\tau_2 = T$.

In view of (3.34),

$$\sum_{t=\tau_1}^{\tau_2-1} u(x_t, x_{t+1}) \geq U(x_{\tau_1}, x_{\tau_2}, \tau_1, \tau_2) - \delta.$$

Combined with (3.37), property (P4), (3.33), and the inequality $T > 2\tau$ this implies that

$$\rho(x_t, \bar{x}) \leq \epsilon, \ t = \tau_1, \ldots, \tau_2.$$

This completes the proof of Theorem 3.2.

3.4 Proof of Theorem 3.3

Theorem 3.2 implies that there exist a positive number δ and an integer $\tau_0 > 0$ such that the following property holds:

(P6) For every natural number $T > 2\tau_0$ and every program $\{x_t\}_{t=0}^T$ which satisfies

$$\sum_{t=0}^{T-1} u(x_t, x_{t+1}) \geq U(x_0, x_T, 0, T) - \delta,$$

there exist integers $\tau_1 \in [0, \tau_0]$, $\tau_2 \in [T - \tau_0, T]$ such that

$$\rho(x_t, \bar{x}) \leq \epsilon, \ t = \tau_1, \ldots, \tau_2.$$

Choose an integer $\tau > 0$ satisfying

$$\tau > 8\tau_0(\delta^{-1}M + 1). \tag{3.38}$$

Set $M_1 = \tau$.

Assume that an integer $T > \tau$ and that a program $\{x_t\}_{t=0}^T$ satisfies

$$\sum_{t=0}^{T-1} u(x_t, x_{t+1}) \geq U(x_0, x_T, 0, T) - M. \tag{3.39}$$

Set $T_0 = 0$. By induction we define a strictly increasing sequence of natural numbers T_i, $i = 0, 1, \ldots$ as follows.

Assume that $T_0, \ldots, T_k$ have been defined where an integer $k \geq 0$. If $T_k = T$, then the construction of the sequence is completed. Assume that $T_k < T$. If $T \leq T_k + 2\tau_0$, then set $T_{k+1} = T$ and the construction is completed. Assume that $T_k < T - 2\tau_0$. If

$$\sum_{t=T_k}^{T-1} u(x_t, x_{t+1}) \geq U(x_{T_k}, x_T, 0, T - T_k) - \delta,$$

then set $T_{k+1} = T$ and the construction is completed. Assume that

$$\sum_{t=T_k}^{T-1} u(x_t, x_{t+1}) < U(x_{T_k}, x_T, 0, T - T_k) - \delta.$$

Then there exists an integer $T_{k+1} > T_k$ such that $T_{k+1} \leq T$ and that

$$\sum_{t=T_k}^{T_{k+1}-1} u(x_t, x_{t+1}) < U(x_{T_k}, x_{T_{k+1}}, 0, T_{k+1} - T_k) - \delta, \tag{3.40}$$

$$\sum_{t=T_k}^{T_{k+1}-2} u(x_t, x_{t+1}) \geq U(x_{T_k}, x_{T_{k+1}-1}, 0, T_{k+1} - T_k - 1) - \delta. \tag{3.41}$$

Therefore by induction we have constructed a finite sequence $T_i, i = 0, \ldots, q$ where q is a natural number and $T_q = T$ is the last element of the sequence. It follows from (3.39) and the construction of $\{T_k\}_{k=0}^{q}$ (see (3.40), (3.41)) that

$$M \geq U(x_0, x_T, 0, T) - \sum_{t=0}^{T-1} u(x_t, x_{t+1})$$

$$\geq \sum \left\{ U(x_{T_k}, x_{T_{k+1}}, 0, T_{k+1} - T_k) - \sum_{t=T_k}^{T_{k+1}-1} u(x_t, x_{t+1}) : \right.$$

$$\left. k \text{ is an integer such that } 0 \leq k < q - 1 \right\} \geq \delta |q - 1|,$$

$$|q - 1| \leq \delta^{-1} M \text{ and } q \leq \delta^{-1} M + 1. \tag{3.42}$$

It follows from the construction of $T_i, i = 0, \ldots, q$ that if an integer k satisfies $0 \leq k \leq q - 1$ and $T_{k+1} - T_k - 1 > 2\tau_0$, then (3.41) is valid. Combined with property (P6) this implies that

$$\rho(x_t, \bar{x}) \leq \epsilon, \ t = T_k + \tau_0, \ldots, T_{k+1} - 1 - \tau_0.$$

Together with (3.42) and (3.38) this implies that

$$\text{Card}(\{t \in \{0, \ldots, T\} : \rho(x_t, \bar{x}) > \epsilon\})$$

$$\leq \text{Card}(\{0, \ldots, T\} \setminus \cup(\{[T_i + \tau_0, \ldots, T_{i+1} - 1 - \tau_0] : i \text{ is an integer such that}$$

$$0 \leq i \leq q - 1, \ T_{i+1} - T_i - 1 > 2\tau_0\}))$$

$$\leq 2q\tau_0 + 4q\tau_0 < 8q\tau_0 \leq 8\tau_0(\delta^{-1}M + 1) < \tau = M_1.$$

This completes the proof of Theorem 3.3.

3.5 Proof of Theorem 3.4

Proposition 2.2 implies that for every natural number T there exists a program $\{x_t^{(T)}\}_{t=0}^{T}$ which satisfies

$$x_0^{(T)} = z, \ \sum_{t=0}^{T-1} u\left(x_t^{(T)}, x_{t+1}^{(T)}\right) = U(z, 0, T). \tag{3.43}$$

It follows from Proposition 3.7 and (3.43) that for every integer $T > 0$ and every pair of integers S_1, S_2 satisfying $0 \leq S_1 < S_2 \leq T$ and $S_2 - S_1 > 2\bar{L}$, we have

$$\sum_{t=S_1}^{S_2-1} u\left(x_t^{(T)}, x_{t+1}^{(T)}\right) \geq (S_2 - S_1)\mu - 4(\bar{L} + 1)(\|u\|). \tag{3.44}$$

Theorem 3.2 and (3.43) imply that the following property holds:

(P7) For every positive number ϵ there exists an integer $\tau_\epsilon > 0$ such that for every natural number $T > 2\tau_\epsilon$, we have

$$\rho\left(x_t^{(T)}, \bar{x}\right) \leq \epsilon, \ t = \tau_\epsilon, \ldots, T - \tau_\epsilon.$$

It is clear that there exists a strictly increasing sequence of natural numbers $\{T_j\}_{j=1}^{\infty}$ such that for all integers $t \geq 0$ there exists

$$x_t = \lim_{j \to \infty} x_t^{(T_j)}. \tag{3.45}$$

Evidently, $\{x_t\}_{t=0}^{\infty}$ is a program satisfying

$$x_0 = z. \tag{3.46}$$

In view of (3.44) and (3.45), for all pairs of integers S_1, S_2 satisfying $S_1 < S_2$ and $S_2 - S_1 > 2\bar{L}$ the inequality

$$\sum_{t=S_1}^{S_2-1} u(x_t, x_{t+1}) \geq (S_2 - S_1)\mu - 4\left(\bar{L} + 1\right)(\|u\|) \tag{3.47}$$

is valid. Proposition 3.1, assumption (B1), and (3.47) imply that $\{x_t\}_{t=0}^\infty$ is a good program and that

$$\lim_{t\to\infty} x_t = \bar{x}. \tag{3.48}$$

We claim that the program $\{x_t\}_{t=0}^\infty$ satisfies (3.9) for every program $\{y_t\}_{t=0}^\infty$ such that $y_0 = z$.

Assume the contrary. Then there exists a program $\{y_t\}_{t=0}^\infty$ such that

$$y_0 = z, \ \gamma := \limsup_{T\to\infty} \sum_{t=0}^{T-1} [u(y_t, y_{t+1}) - u(x_t, x_{t+1})] > 0. \tag{3.49}$$

In view of (3.49), Proposition 3.1, (3.47), and assumption (B1) $\{y_t\}_{t=0}^\infty$ is a good program and

$$\lim_{t\to\infty} y_t = \bar{x}. \tag{3.50}$$

Assumption (B2) implies that there exists a number $\epsilon \in (0, 1)$ such that for every point $(y_1, y_2) \in \Omega$ which satisfy

$$\rho(y_1, \bar{x}), \ \rho(y_2, \bar{x}) \leq \epsilon$$

the inequality

$$|u(y_1, y_2) - u(\bar{x}, \bar{x})| \leq \gamma(16L_*)^{-1}$$

holds.

Assumption (B3) implies that there exists a number $\delta \in (0, \epsilon)$ such that the following property holds:

(P8)   For every pair of points $y_1, y_2 \in K$ which satisfies $\rho(y_i, \bar{x}) \leq \delta$, $i = 1, 2$ there exists a program $\{z_t\}_{t=0}^{L_*}$ such that

$$z_0 = y_1, \ z_{L_*} = y_2, \ \rho(y_t, \bar{x}) \leq \epsilon, \ t = 0, \ldots, L_*.$$

It follows from (3.48) and (3.50) that there exists an integer $S_0 > 0$ such that

$$\rho(x_t, \bar{x}), \ \rho(y_t, \bar{x}) \leq \delta/4 \text{ for all integers } t \geq S_0. \tag{3.51}$$

In view of (3.48), there exists an integer $S_1 > S_0 + 4L_* + 4$ for which

$$\sum_{t=0}^{S_1-1} [u(y_t, y_{t+1}) - u(x_t, x_{t+1})] > \gamma/2. \tag{3.52}$$

It follows from (3.45) and upper semicontinuity of the function u that there exists an integer number $k > 0$ such that

$$T_k > S_1 + 4, \tag{3.53}$$

$$\rho\left(x_t^{(T_k)}, x_t\right) \le \delta/4, \ t = 0, \ldots, S_1 + 4, \tag{3.54}$$

$$u\left(x_t^{(T_k)}, x_{t+1}^{(T_k)}\right) \le u(x_t, x_{t+1}) + \gamma(16(S_1 + 4))^{-1}, \ t = 0, \ldots, S_1 + 4. \tag{3.55}$$

Relations (3.51), (3.53), and (3.54) imply that

$$\rho(y_{S_1-L_*}, \bar{x}) \le \delta/4, \ \rho\left(x_{S_1}^{(T_k)}, \bar{x}\right) \le \rho\left(x_{S_1}^{(T_k)}, x_{S_1}\right) + \rho(x_{S_1}, \bar{x}) \le \delta/4 + \delta/4. \tag{3.56}$$

By (3.56) and property (P8), there exists a program $\{\bar{x}_t\}_{t=0}^{T_k}$ such that

$$\bar{x}_t = y_t, \ t \in \{0, \ldots, S_1 - L_*\}, \ \bar{x}_t = x_t^{(T_k)}, \ t = S_1, \ldots, T_k, \tag{3.57}$$

$$\rho(\bar{x}_t, \bar{x}) \le \epsilon, \ t = S_1 - L_*, \ldots, S_1.$$

It follows from (3.57), (3.49), (3.43), (3.55), the choice of ϵ, (3.51), and (3.52) that

$$0 \le \sum_{t=0}^{T_k-1} u(x_t^{(T_k)}, x_{t+1}^{(T_k)}) - \sum_{t=0}^{T_k-1} u(\bar{x}_t, \bar{x}_{t+1})$$

$$= \sum_{t=0}^{S_1-1} u(x_t^{(T_k)}, x_{t+1}^{(T_k)}) - \sum_{t=0}^{S_1-1} u(\bar{x}_t, \bar{x}_{t+1}) \le \sum_{t=0}^{S_1-1} u(x_t, x_{t+1}) + \gamma(16)^{-1}$$

$$- \sum_{t=0}^{S_1-L_*-1} u(y_t, y_{t+1}) - (u(\bar{x}, \bar{x}) - \gamma(16L_*)^{-1})(2L_*)$$

$$\le \gamma(16)^{-1} + \gamma/8 + \sum_{t=0}^{S_1-1} u(x_t, x_{t+1})$$

$$- \sum_{t=0}^{S_1-1} u(y_t, y_{t+1}) + \gamma/8 \le \gamma/4 + \gamma 16^{-1} - \gamma/2 < 0.$$

The contradiction we have reached completes the proof of Theorem 3.4.

3.6   Proof of Theorem 3.5

It is clear that if $\{x_t\}_{t=0}^{\infty}$ is an overtaking optimal program, then it is locally optimal. Assume that the program $\{x_t\}_{t=0}^{\infty}$ is locally optimal. Proposition 3.7 implies that the program $\{x_t\}_{t=0}^{\infty}$ is good. By assumption (B1),

$$\lim_{t \to \infty} x_t = \bar{x}. \tag{3.58}$$

Theorem 3.4 implies that there exists an overtaking optimal program $\{y_t\}_{t=0}^{\infty}$ satisfying

$$y_0 = x_0. \tag{3.59}$$

It is clear that the program $\{y_t\}_{t=0}^{\infty}$ is good. In view of assumption (B1),

$$\lim_{t \to \infty} y_t = \bar{x}. \tag{3.60}$$

We show that the program $\{x_t\}_{t=0}^{\infty}$ is overtaking optimal. Assume the contrary. Then

$$\gamma := \limsup_{T \to \infty} \left[\sum_{t=0}^{T-1} u(y_t, y_{t+1}) - u(x_t, x_{t+1}) \right] > 0. \tag{3.61}$$

Assumption (B2) implies that there exists a number $\epsilon \in (0, \gamma/2)$ such that for every point $(z_1, z_2) \in \Omega$ which satisfies

$$\rho(z_1, \bar{x}), \rho(z_2, \bar{x}) \le \epsilon,$$

the inequality

$$|u(z_1, z_2) - u(\bar{x}, \bar{x})| \le \gamma (16 L_*)^{-1} \tag{3.62}$$

holds.

In view of assumption (B3), there exists a number $\delta \in (0, \epsilon)$ such that the following property holds:

(P9)   For every pair of points $z_1, z_2 \in K$ which satisfies $\rho(z_i, \bar{x}) \le \delta, i = 1, 2$ there exists a program $\{\xi_t\}_{t=0}^{L_*}$ such that

$$\xi_0 = z_1, \ \xi_{L_*} = z_2, \ \rho(\xi_t, \bar{x}) \le \epsilon, \ t = 0, \dots, L_*.$$

It follows from (3.58) and (3.60) that there exists an integer $S_0 > 4L_* + 4$ such that

$$\rho(x_t, \bar{x}), \ \rho(y_t, \bar{x}) \le \delta \text{ for all integers } t \ge S_0 - L_* - 2. \tag{3.63}$$

In view of (3.61), there exists a natural number $S_1 > S_0 + 2 + 2L_*$ such that

$$\sum_{t=0}^{S_1-1} (u(y_t, y_{t+1}) - u(x_t, x_{t+1})) > \gamma/2. \tag{3.64}$$

Property (P9) and (3.63) imply that there exists a program $\{\tilde{x}_t\}_{t=0}^{S_1}$ such that

$$\tilde{x}_t = y_t, \ t = 0, \ldots, S_1 - L_*, \ \rho(x_t, \tilde{x}) \leq \epsilon, \ t = S_1 - L_*, \ldots, S_1, \tag{3.65}$$

$$\tilde{x}_{S_1} = x_{S_1}.$$

It follows from local optimality of $\{x_t\}_{t=0}^{\infty}$, (3.65), (3.59), (3.64), (3.63), and the choice of ϵ (see (3.62)) that

$$0 \leq \sum_{t=0}^{S_1-1} u(x_t, x_{t+1}) - \sum_{t=0}^{S_1-1} u(\tilde{x}_t, \tilde{x}_{t+1})$$

$$= \sum_{t=0}^{S_1-1} u(x_t, x_{t+1}) - \sum_{t=0}^{S_1-1} u(y_t, y_{t+1})$$

$$+ \sum_{t=S_1-L_*}^{S_1-1} u(y_t, y_{t+1}) - \sum_{t=S_1-L_*}^{S_1-1} u(\tilde{x}_t, \tilde{x}_{t+1})$$

$$\leq -\gamma/2 + \gamma 16^{-1} L_*^{-1} (2L_* + 2) < 0,$$

a contradiction. The contraction we have reached proves Theorem 3.5.

3.7 Proof of Theorem 3.6

In view of Theorem 3.2, there exist an integer $\tau_0 > 0$ and a positive number δ such that the following property holds:

(P10)   For every natural number $T > 2\tau_0$ and every program $\{x_t\}_{t=0}^{T}$ which satisfies

$$\sum_{t=0}^{T-1} u(x_t, x_{t+1}) \geq U(x_0, x_T, 0, T) - \delta,$$

there exist integers $\tau_1 \in [0, \tau_0]$, $\tau_2 \in [T - \tau_0, T]$ such that

$$\rho(x_t, \bar{x}) \leq \epsilon, \ t = \tau_1, \ldots, \tau_2,$$

$$\tau_1 = 0 \text{ if } \rho(x_0, \bar{x}) \leq \delta \text{ and } \tau_2 = T \text{ if } \rho(x_T, \bar{x}) \leq \delta.$$

Set

$$\tau = 3\tau_0 + 1. \tag{3.66}$$

Assume that an integer $T > 2\tau$ and that a program $\{x_t\}_{t=0}^T$ for every integer $S \in \{0, \ldots, T - \tau\}$ satisfies

$$\sum_{t=S}^{S+\tau-1} u(x_t, x_{t+1}) \geq U(x(S), x(S + \tau), S, S + \tau) - \delta. \tag{3.67}$$

Property (P10), (3.66), and (3.67) imply that

$$\rho(x_t, \bar{x}) \leq \epsilon, \ t = \tau_0, \ldots, \tau - \tau_0,$$

$$\rho(x_t, \bar{x}) \leq \epsilon, \ t = T - \tau + \tau_0, \ldots, T - \tau_0,$$

if $\rho(x_0, \bar{x}) \leq \delta$, then $\rho(x_t, \bar{x}) \leq \epsilon, t = 0, \ldots, \tau - \tau_0$ and if $\rho(x_T, \bar{x}) \leq \delta$, then

$$\rho(x_t, \bar{x}) \leq \epsilon, \ t = T - \tau + \tau_0, \ldots, T.$$

In order to complete the proof it is sufficient to show that

$$\rho(x_t, \bar{x}) \leq \epsilon, \ t = \tau - \tau_0, \ldots, T - \tau_0.$$

Assume that an integer

$$S \in [\tau - \tau_0, T - \tau_0]. \tag{3.68}$$

It follows from (3.66) and (3.68) that there exist integers $S_1, S_2 \in [0, T]$ such that

$$S_1 + \tau_0 \leq S \leq S_2 - \tau_0, \ S_2 - S_1 = \tau. \tag{3.69}$$

In view of (3.69), (3.67), and property (P10),

$$\rho(x_S, \bar{x}) \leq \epsilon.$$

Theorem 3.6 is proved.

3.8 Stability Results

For every function $\phi : \Omega \to R^1$ set

$$\|\phi\| = \sup\{|\phi(z)| \; z \in \Omega\}.$$

Let T_1, T_2 be integers such that $T_1 < T_2$ and $u_t : \Omega \to R^1, t = T_1, \ldots T_2 - 1$ be bounded functions. For every pair of points $z_0, z_1 \in K$ define

$$U\left(\{u_t\}_{t=T_1}^{T_2-1}, z_0, z_1\right) = \sup \left\{ \sum_{t=T_1}^{T_2-1} u_t(x_t, x_{t+1}) : \{x_t\}_{t=T_1}^{T_2} \text{ is a program} \right.$$

$$\left. \text{such that } x_{T_1} = z_0, \; x_{T_2} = z_1 \right\}$$

and

$$U\left(\{u_t\}_{t=T_1}^{T_2-1}, z_0\right) = \sup \left\{ \sum_{t=T_1}^{T_2-1} u_t(x_t, x_{t+1}) : \{x_t\}_{t=T_1}^{T_2} \text{ is a program} \right.$$

$$\left. \text{such that } x_{T_1} = z_0 \right\}.$$

(Here we assume that supremum over an empty set is $-\infty$.) It is easy to see that the following result holds.

Proposition 3.11 *Let* $y_0, \tilde{y}_0 \in K$, $T_1 < T_2$ *be integers,* u_t, $t = T_1, \ldots, T_2 - 1$ *be bounded and upper semicontinuous functions and let* $U(\{u_t\}_{t=T_1}^{T_2-1}, y_0, \tilde{y}_0)$ *be finite. Then there exists a program* $\{x_t\}_{t=T_1}^{T_2}$ *such that*

$$\sum_{t=T_1}^{T_2-1} u_t(x_t, x_{t+1}) = U\left(\{u_t\}_{t=T_1}^{T_2-1}, y_0, \tilde{y}_0\right), \quad x_{T_1} = y_0, \; x_{T_2} = \tilde{y}_0.$$

Theorem 3.12 *Let* ϵ *be a positive number. Then there exist an integer* $\tau > 0$ *and a positive number* γ *such that for each natural number* $T > 2\tau$, *each sequence of bounded functions* $u_t : \Omega \to R^1, t = 0, \ldots, T - 1$ *satisfying* $\|u_t - u\| \leq \gamma$, $t = 0, 1, \ldots, T - 1$ *and each program* $\{x_t\}_{t=0}^{T}$ *which satisfies*

$$\sum_{t=S}^{S+\tau-1} u_t(x_t, x_{t+1}) \geq U\left(\{u_t\}_{t=S}^{S+\tau-1}, x_S, x_{S+\tau}\right) - \gamma$$

for every integer $S \in \{0, \ldots, T - \tau\}$ there exist integers τ_1, τ_2 such that

$$\tau_1 \in [0, \tau], \quad \tau_2 \in [T - \tau, T],$$

$$\rho(x_t, \bar{x}) \leq \epsilon, \quad t = \tau_1, \ldots \tau_2.$$

Moreover, $\tau_1 = 0$ if $\rho(x_0, \bar{x}) \leq \gamma$ and $\tau_2 = T$ if $\rho(x_T, \bar{x}) \leq \gamma$.

Proof Theorem 3.12 follows easily from Theorem 3.6. Namely let a natural number τ and $\delta > 0$ be as guaranteed by Theorem 3.6. Put

$$\gamma = \delta(4(\tau + 1))^{-1}.$$

Now it is not difficult to see that the assertion of Theorem 3.12 holds.

Theorem 3.12 implies the following result.

Theorem 3.13 *Let $\epsilon > 0$. Then there exist a natural number τ, $\lambda > 1$, and $\gamma > 0$ such that for each integer $T > 2\tau$, each sequence of bounded functions $u_t : \Omega \rightarrow R^1$, $t = 0, \ldots, T - 1$ satisfying $\|u_t - u\| \leq \gamma$, $t = 0, 1, \ldots, T - 1$, each sequence $\{\alpha_i\}_{i=0}^{T-1} \subset (0, 1]$ such that for each $i, j \in \{0, \ldots, T - 1\}$ satisfying $|i - j| \leq \tau$ the inequality $\alpha_i \alpha_j^{-1} \leq \lambda$ holds and each program $\{x_t\}_{t=0}^T$ which satisfies*

$$\sum_{t=S}^{S+\tau-1} \alpha_t u_t(x_t, x_{t+1}) \geq U\left(\{\alpha_t u_t\}_{t=S}^{S+\tau-1}, x_S, x_{S+\tau}\right) - \gamma \alpha_S$$

for each integer $S \in \{0, \ldots, T - \tau\}$ there exist integers τ_1, τ_2 such that

$$\tau_1 \in [0, \tau], \quad \tau_2 \in [T - \tau, T], \tag{3.70}$$

$$\rho(x_t, \bar{x}) \leq \epsilon, \quad t = \tau_1, \ldots \tau_2.$$

Moreover,

$$\tau_1 = 0 \text{ if } \rho(x_0, \bar{x}) \leq \gamma, \tag{3.71}$$

and

$$\tau_2 = T \text{ if } \rho(x_T, \bar{x}) \leq \gamma. \tag{3.72}$$

Theorem 3.13 implies the following result.

Theorem 3.14 *Let $\epsilon > 0$. Then there exist a natural number τ, $\gamma > 0$, and $\lambda > 1$ such that for each integer $T > 2\tau$, each sequence of bounded upper semicontinuous functions $u_t : \Omega \rightarrow R^1$, $t = 0, \ldots, T-1$ satisfying $\|u_t - u\| \leq \gamma$, $t = 0, 1, \ldots, T-$*

1, *each sequence* $\{\alpha_i\}_{i=0}^{T-1} \subset (0, 1]$ *such that for each* $i, j \in \{0, \ldots, T-1\}$ *satisfying* $|i-j| \leq \tau$ *the inequality* $\alpha_i \alpha_j^{-1} \leq \lambda$ *holds and each program* $\{x_t\}_{t=0}^{T}$ *which satisfies*

$$\sum_{t=0}^{T-1} \alpha_t u_t(x_t, x_{t+1}) = U\left(\{\alpha_t u_t\}_{t=0}^{T-1}, x_0, x_T\right)$$

there exist integers τ_1, τ_2 *such that (3.70)–(3.72) hold.*

Theorems 3.12–3.14 were obtained from [83]. The next turnpike result is new.

Theorem 3.15 *Let* ϵ, M *be positive numbers. Then there exist a natural number* Q *and* $\gamma > 0$ *such that for each integer* $T > Q$, *each finite sequence of bounded functions* $u_t : \Omega \to R^1, t = 0, \ldots, T-1$ *satisfying* $\|u_t - u\| \leq \gamma$ $t = 0, 1, \ldots, T-1$, *and each program* $\{x_t\}_{t=0}^{T}$ *which satisfies*

$$\sum_{t=0}^{T-1} u_t(x_t, x_{t+1}) \geq U(\{u_t\}_{t=0}^{T-1}, x_0, x_T) - M$$

the following inequality holds:

$$\mathrm{Card}(\{t \in \{0, \ldots, T\} : \rho(x_t, \bar{x}) > \epsilon\}) \leq Q.$$

Proof Theorem 3.12 implies that there exist an integer $\tau \geq 1$ and a positive number γ such that the following property holds:

(i) for each integer $S > 2\tau$, each finite sequence of bounded functions $u_t : \Omega \to R^1, t = 0, \ldots, S-1$ satisfying

$$\|u_t - u\| \leq \gamma, \ t = 0, \ldots, S-1$$

and each program $\{x_t\}_{t=0}^{S}$ which satisfies

$$\sum_{t=0}^{S-1} u_t(x_t, x_{t+1}) \geq U(\{u_t\}_{t=0}^{S-1}, x_0, x_S) - \gamma$$

we have

$$\rho(x_t, \bar{x}) \leq \epsilon, \ t = \tau, \ldots, S - \tau.$$

Choose a natural number

$$Q > 9\tau(\gamma^{-1}M + 1). \tag{3.73}$$

Assume that an integer $T > Q$, a finite sequence of bounded functions $u_t : \Omega \to R^1, t = 0, \ldots, T - 1$ satisfies

$$\|u_t - u\| \leq \gamma, \ t = 0, \ldots, T - 1 \tag{3.74}$$

and that a program $\{x_t\}_{t=0}^T$ satisfies

$$\sum_{t=0}^{T-1} u_t(x_t, x_{t+1}) \geq U\left(\{u_t\}_{t=0}^{S-1}, x_0, x_T\right) - M. \tag{3.75}$$

Set $T_0 = 0$. By induction we define a strictly increasing sequence of integers $T_0 < \cdots < T_k \leq T$.

Assume that $k \geq 0$ is an integer and $T_0, \ldots, T_k$ have been defined. If $T_k = T$, then the construction of the sequence is completed. Assume that $T_k < T$. In $T \leq T_k + 2\tau$, then we set $T_{k+1} = T$ and the construction is completed.

Assume that

$$T_k < T - 2\tau.$$

If

$$\sum_{t=T_k}^{T-1} u_t(x_t, x_{t+1}) \geq U\left(\{u_t\}_{t=T_k}^{T-1}, x_{T_k}, x_T\right) - \gamma,$$

then set $T_{k+1} = T$ and the construction is completed.

Assume that

$$\sum_{t=T_k}^{T-1} u_t(x_t, x_{t+1}) < U\left(\{u_t\}_{t=T_k}^{T-1}, x_{T_k}, x_T\right) - \gamma.$$

Then there exists an integer $T_{k+1} \leq T$ such that

$$T_{k+1} \geq T_k + 2,$$

$$\sum_{t=T_k}^{T_{k+1}-1} u_t(x_t, x_{t+1}) < U\left(\{u_t\}_{t=T_k}^{T_{k+1}-1}, x_{T_k}, x_{T_{k+1}}\right) - \gamma \tag{3.76}$$

and

$$\sum_{t=T_k}^{T_{k+1}-2} u_t(x_t, x_{t+1}) \geq U\left(\{u_t\}_{t=T_k}^{T_{k+1}-2}, x_{T_k}, x_{T_{k+1}-1}\right) - \gamma. \tag{3.77}$$

Thus by induction we have constructed a finite sequence of integers

$$T_0 = 0 < \cdots, T_q = T,$$

where $q > 0$ is an integer. It follows from (3.75) and the construction of $\{T_i\}_{i=0}^q$ (see (3.76) and (3.77)) that

$$M \geq U(\{u_t\}_{t=0}^{T-1}, x_0, x_T) - \sum_{t=0}^{T-1} u_t(x_t, x_{t+1})$$

$$\geq \sum \{U(\{u_t\}_{t=T_k}^{T_{k+1}-1}, x_{T_k}, x_{T_{k+1}}) - \sum_{t=T_k}^{T_{k+1}-1} u_t(x_t, x_{t+1}) :$$

$$k \text{ is an integer satisfying } 0 \leq k < q - 1\} \geq \gamma(q - 1)$$

and

$$q \leq \gamma^{-1} M + 1.$$

It follows from property (i), (3.74), and the construction of $T_i, i = 0, \ldots, q$ that if an integer k satisfies

$$0 \leq k \leq q - 1, \ T_{k+1} - T_k > 2\tau + 1,$$

then (3.77) holds and

$$\rho(x_t, \bar{x}) \leq \epsilon, \ t = T_k + \tau, \ldots, T_{k+1} - \tau - 1.$$

Combined with (3.73) and the inequality

$$q \leq \gamma^{-1} M + 1,$$

this implies that

$$\text{Card}(\{t \in \{0, \ldots, T\} : \rho(x_t, \bar{x}) > \epsilon\})$$

$$\leq \text{Card}(\{t \in \{0, \ldots, T\} \setminus \cup(\{T_i + \tau, \ldots, T_{i+1} - \tau - 1\} :$$

$$i \text{ is an integer satisfying } 0 \leq i \leq q - 1, \ T_{i+1} - T_i > 2\tau + 1)\})$$

$$\leq 3q\tau + 6q\tau \leq 9\tau(\gamma^{-1} M + 1) < Q.$$

Theorem 3.15 is proved.

3.9 Agreeable Programs

A program $\{x_t^*\}_{t=0}^{\infty}$ is called agreeable if for any natural number T_0 and any $\epsilon > 0$ there exists an integer $T_\epsilon > T_0$ such that for any integer $T \geq T_\epsilon$ there exists a program $\{x_t\}_{t=0}^{T}$ which satisfies

$$x_t = x_t^*, \ t = 0, \ldots, T_0$$

and

$$\sum_{t=0}^{T-1} u(x_t, x_{t+1}) \geq U(x_0^*, 0, T) - \epsilon.$$

The notion of agreeable programs is well known in the economic literature [34–36]. In this section we employ a strong version of it.

A program $\{x_t^*\}_{t=0}^{\infty}$ is called strongly agreeable if for any natural number T_0 and any $\epsilon > 0$ there exist an integer $T_\epsilon > T_0$ and $\delta > 0$ such that for each integer $T \geq T_\epsilon$ and each finite sequence of bounded functions $u_t : \Omega \to R^1, \ t = 0, \ldots, T$ satisfying

$$\|u_t - u\| \leq \delta, \ t = 0 \ldots, T - 1$$

there exists a program $\{x_t\}_{t=0}^{T}$ which satisfies

$$x_t = x_t^*, \ t = 0, \ldots, T_0$$

and

$$\sum_{t=0}^{T-1} u_t(x_t, x_{t+1}) \geq U\left(\{u_t\}_{t=0}^{T-1}, x_0^*\right) - \epsilon.$$

Theorem 3.16 *Let* $\{x_t\}_{t=0}^{\infty}$ *be a program. Then the following properties are equivalent:*

(a) the program $\{x_t\}_{t=0}^{\infty}$ *is locally optimal;*
(b) the program $\{x_t\}_{t=0}^{\infty}$ *is agreeable;*
(c) the program $\{x_t\}_{t=0}^{\infty}$ *is strongly agreeable.*

Proof Clearly, property (c) implies property (b) and property (b) implies property (a). We show that (a) implies (c).

Assume that the program $\{x_t\}_{t=0}^{\infty}$ is locally optimal. We show that the program $\{x_t\}_{t=0}^{\infty}$ is strongly agreeable.

Let $\epsilon \in (0, 1)$ and T_0 be a natural number. By (B2) there is a positive number $\delta_1 < \epsilon/2$ such that the following property holds:

(i) for each $(x, y) \in \Omega$ satisfying

$$\rho(x, \bar{x}) \le \delta_1, \ \rho(y, \bar{x}) \le \delta_1$$

we have

$$|u(x, y) - u(\bar{x}, \bar{x})| \le (\epsilon/8)(L_* + 1)^{-1}.$$

By (B3) there exists $\delta_2 \in (0, \delta_1)$ such that the following property holds:
(ii) for each $z_1, z_2 \in K$ satisfying

$$\rho(z_i, \bar{x}) \le \delta_2, \ i = 1, 2$$

there exists a program $\{\xi_t\}_{t=0}^{L_*}$ such that

$$\rho(\xi_t, \bar{x}) \le \delta_1, \ t = 0, \ldots, L_*,$$

$$\xi_0 = z_1, \ \xi_{L_*} = z_2.$$

By Theorem 3.12, there exist a natural number L_1 and $\delta_3 \in (0, \delta_2)$ such that the following property holds:
(iii) for each integer $T > 2L_1$, each finite sequence of bounded functions $u_t : \Omega \to R^1$, $t = 0, \ldots, T - 1$ satisfying

$$\|u_t - u\| \le \delta_3, \ t = 0 \ldots, T - 1$$

and each program $\{z_t\}_{t=0}^{T}$ which satisfies

$$\sum_{t=0}^{T-1} u_t(z_t, z_{t+1}) \ge U(\{u_t\}_{t=0}^{T-1}, z_0, z_T) - \delta_3$$

we have

$$\rho(z_t, \bar{x}) \le \delta_2, \ t = L_1, \ldots, T - L_1.$$

Properties (a) and (iii) imply that

$$\rho(x_t, \bar{x}) \le \delta_2 \text{ for all integers } t \ge L_1. \tag{3.78}$$

Choose an integer

$$T_\epsilon > T_0 + 2L_* + 2L_1 + 8 \tag{3.79}$$

and set

$$\delta = \delta_3 (2T_\epsilon + 2)^{-1}/16. \tag{3.80}$$

Assume that an integer $T \geq T_\epsilon$, a finite sequence of bounded functions $u_t : \Omega \to R^1$, $t = 0, \ldots, T - 1$ satisfies

$$\|u_t - u\| \leq \delta, \ t = 0 \ldots, T - 1. \tag{3.81}$$

There exists a program $\{y_t\}_{t=0}^T$ such that

$$y_0 = x_0, \tag{3.82}$$

$$\sum_{t=0}^{T-1} u_t(y_t, y_{t+1}) \geq U\left(\{u_t\}_{t=0}^{T-1}, x_0\right) - \delta. \tag{3.83}$$

It follows from property (iii) and (3.79)–(3.83) that

$$\rho(y_t, \bar{x}) \leq \delta_2, \ t = L_1, \ldots, T - L_1. \tag{3.84}$$

Property (ii), (3.78), (3.79), and (3.84) imply that there exists a program $\{z_t\}_{t=0}^{T_\epsilon}$ such that

$$z_t = y_t, \ t = 0, \ldots, T_\epsilon - L_* - L_1, \tag{3.85}$$

$$z_{T_\epsilon - L_1} = x_{T_\epsilon - L_1}, \tag{3.86}$$

$$\rho(z_t, \bar{x}) \leq \delta_1, \ t = T_\epsilon - L_1 - L_*, \ldots, T_\epsilon - L_1, \tag{3.87}$$

$$z_t = x_t, \ t = T_\epsilon - L_1, \ldots, T_\epsilon. \tag{3.88}$$

By properties (a) and (i) and (3.81), (3.85)–(3.87),

$$\sum_{t=0}^{T_\epsilon - L_1 - 1} u(x_t, x_{t+1}) \geq \sum_{t=0}^{T_\epsilon - L_1 - 1} u(z_t, z_{t+1})$$

$$\geq \sum_{t=0}^{T_\epsilon - L_* - L_1 - 1} u(y_t, y_{t+1}) + L_*(u(\bar{x}, \bar{x})$$

$$- (\epsilon/8)(L_* + 1)^{-1})$$

$$\geq \sum_{t=0}^{T_\epsilon - L_* - L_1 - 1} u(y_t, y_{t+1}) + L_* u(\bar{x}, \bar{x}) - \epsilon/8. \tag{3.89}$$

Property (i), (3.78), (3.79), and (3.89) imply that

$$\sum_{t=0}^{T_\epsilon-L_*-L_1-1} u(x_t, x_{t+1})$$

$$\geq \sum_{t=0}^{T_\epsilon-L_*-L_1-1} u(y_t, y_{t+1}) + L_* u(\bar{x}, \bar{x}) - \epsilon/8$$

$$- \sum_{t=T_\epsilon-L_*-L_1-1}^{T_\epsilon-L_1-1} u(x_t, x_{t+1})$$

$$\geq \sum_{t=0}^{T_\epsilon-L_*-L_1-1} u(y_t, y_{t+1}) + L_* u(\bar{x}, \bar{x}) - \epsilon/8 - L_1 u(\bar{x}, \bar{x}) - \epsilon/8$$

$$= \sum_{t=0}^{T_\epsilon-L_*-L_1-1} u(y_t, y_{t+1}) - \epsilon/4. \qquad (3.90)$$

It follows from (3.80), (3.81), and (3.90) that

$$\sum_{t=0}^{T_\epsilon-L_*-L_1-1} u_t(x_t, x_{t+1}) \geq \sum_{t=0}^{T_\epsilon-L_*-L_1-1} u_t(y_t, y_{t+1}) - \epsilon/4 - \epsilon/8. \qquad (3.91)$$

Property (ii), (3.78), (3.79), and (3.84) imply that there exists a program $\{\xi_t\}_{t=0}^{T}$ such that

$$\xi_t = x_t, \ t = 0, \dots, T_\epsilon - L_* - L_1, \qquad (3.92)$$

$$\rho(\xi_t, \bar{x}) \leq \delta_1, \ t = T_\epsilon - L_1 - L_* + 1, \dots, T_\epsilon - L_1, \qquad (3.93)$$

$$\xi_t = y_t, \ t = T_\epsilon - L_1, \dots, T. \qquad (3.94)$$

By property (i), (3.78), (3.79), (3.84), and (3.91)–(3.94),

$$\sum_{t=0}^{T-1} u_t(\xi_t, \xi_{t+1}) - \sum_{t=0}^{T-1} u_t(y_t, y_{t+1})$$

$$\geq \sum_{t=0}^{T_\epsilon-L_*-L_1-1} u_t(x_t, x_{t+1}) - \sum_{t=0}^{T_\epsilon-L_*-L_1-1} u_t(y_t, y_{t+1})$$

$$+ \sum_{t=T_\epsilon-L_*-L_1}^{T_\epsilon-L_1-1} u_t(\xi_t, \xi_{t+1}) - \sum_{t=T_\epsilon-L_*-L_1}^{T_\epsilon-L_1-1} u_t(y_t, y_{t+1})$$

$$\geq -3\epsilon/8 - 2L_*(\epsilon/8)(L_* + 1)^{-1} \geq -5\epsilon/8.$$

Combined with (3.80) and (3.83) this implies that

$$\sum_{t=0}^{T-1} u_t(\xi_t, \xi_{t+1}) \geq \sum_{t=0}^{T-1} u_t(y_t, y_{t+1}) - 5\epsilon/8$$

$$\geq U(\{u_t\}_{t=0}^{T-1}, x_0) - 5\epsilon/8 - \delta$$

$$\geq U(\{u_t\}_{t=0}^{T-1}, x_0) - \epsilon.$$

Thus (c) holds. Theorem 3.16 is proved.

Chapter 4
Generic Turnpike Properties

In this chapter we consider a class of discrete-time optimal control problems identified with a complete metric space of objective functions. Using the Baire category approach, we show that for most problems the results of Chap. 3 are true. In particular, a typical (generic) problem possesses the turnpike property.

4.1 Preliminaries

We continue to study the class of optimal control problems introduced in Sect. 2.2 using the same notation, definitions, and assumptions.

Let (K, ρ) be a compact metric space and let Ω be a nonempty closed subset of $K \times K$.

We recall that a sequence $\{x_t\}_{t=0}^{\infty} \subset K$ is called an (Ω)-program (or a program if Ω is understood) if $(x_t, x_{t+1}) \in \Omega$ for all integers $t \geq 0$.

Let the integers T_1, T_2 satisfy $T_1 < T_2$. A sequence $\{x_t\}_{t=T_1}^{T_2} \subset K$ is called an (Ω)-program (or a program if Ω is understood) if $(x_t, x_{t+1}) \in \Omega$ for all integers t satisfying $T_1 \leq t < T_2$.

Let $u : \Omega \to R^1$ be a bounded upper semicontinuous function. Set

$$\|u\| = \sup\{|u(z)| : z \in \Omega\}.$$

Recall that the supremum over empty set is $-\infty$.

For each pair of integers T_1, T_2 satisfying $0 \leq T_1 < T_2$ and each $y, \tilde{y} \in K$ define

$$U(u, y, T_1, T_2) = \sup \left\{ \sum_{t=T_1}^{T_2-1} u(x_t, x_{t+1}) : \{x_t\}_{t=T_1}^{T_2} \text{ is a program and } x_{T_1} = y \right\},$$

A. J. Zaslavski, *Optimal Control Problems Arising in Forest Management*,
SpringerBriefs in Optimization, https://doi.org/10.1007/978-3-030-23587-1_4

$$U(u, y, \tilde{y}, T_1, T_2) = \sup \left\{ \sum_{t=T_1}^{T_2-1} u(x_t, x_{t+1}) : \right.$$

$$\left. \{x_t\}_{t=T_1}^{T_2} \text{ is a program and } x_{T_1} = y, \ x_{T_2} = \tilde{y} \right\},$$

$$\widehat{U}(u, T_1, T_2) = \sup \left\{ \sum_{t=T_1}^{T_2-1} u(x_t, x_{t+1}) : \ \{x_t\}_{t=T_1}^{T_2} \text{ is a program} \right\}.$$

In this chapter we suppose that the following assumption, introduced in Sect. 2.2, holds.

(A)　　There exists a natural number $\bar{L}$ such that for each $y, z \in K$ there is a program $\{x_t\}_{t=0}^{\bar{L}}$ such that $x_0 = y$ and $x_{\bar{L}} = z$.

This assumption holds for the important class of forest management problems.

Note that in Chaps. 2 and 3 the function u was fixed and for each pair of integers T_1, T_2 satisfying $0 \leq T_1 < T_2$ and each $y, \tilde{y} \in K$ we used the notation

$$U(y, \tilde{y}, T_1, T_2) = U(u, y, \tilde{y}, T_1, T_2),$$

$$U(y, T_1, T_2) = U(u, y, T_1, T_2),$$

$$\widehat{U}(T_1, T_2) = U(u, T_1, T_2).$$

Note that all the results stated in Sect. 2.2 (Theorems 2.6, 2.7, 2.9, and 2.10 and Propositions 2.5 and 2.8) hold in our case with $u_t = u$, $t = 0, 1, \ldots$.

The results of this chapter are new.

4.2　Equivalence of the Turnpike Properties

We use the constant μ defined by Theorem 2.9 and suppose that

$$\mu = \sup\{u(x, x) : \ x \in K \text{ and } (x, x) \in \Omega\}. \tag{4.1}$$

By the upper semicontinuity of u, there exists $\bar{x} \in K$ such that

$$(\bar{x}, \bar{x}) \in \Omega \text{ and } u(\bar{x}, \bar{x}) = \mu. \tag{4.2}$$

Recall that a program $\{x_t\}_{t=0}^{\infty}$ is called (u)-good (or good if the function u is understood) if the sequence

$$\left\{ \sum_{t=0}^{T-1} \{u(x_t, x_{t+1}) - \mu\} \right\}_{T=1}^{\infty}$$

is bounded.

A program $\{x_t\}_{t=0}^{\infty}$ is called (u)-bad (or bad if the function u is understood) if $\sum_{t=0}^{T-1}(u(x_t, x_{t+1}) - \mu) \to -\infty$ as $T \to \infty$.

By Proposition 3.1, any program is either good or bad and for any $x_0 \in K$ there is a good program $\{x_t\}_{t=0}^{\infty}$.

We suppose that assumptions (B2) and (B3) introduced in Sect. 3.1 hold.

(B2) u is continuous at $(\bar{x}, \bar{x})$ with respect to Ω.

(B3) There exists a natural number L_* such that for each $\epsilon > 0$ there exists $\delta > 0$ such that for each $y_1, y_2 \in K$ satisfying $\rho(y_1, \bar{x})$, $\rho(\bar{x}, y_2) \le \delta$ there is a program $\{x_t\}_{t=0}^{L_*}$ such that $x_0 = y_1$, $x_{L_*} = y_2$ and $\rho(x_t, \bar{x}) \le \epsilon$, $t = 0, \ldots, L_*$.

In Chap. 3 we also use the following assumption.

(B1) For each good program $\{x_t\}_{t=0}^{\infty}$, $\lim_{t \to \infty} x_t = \bar{x}$.

Property (B1) is known in the literature and is called as the asymptotic turnpike property. Clearly, if (B1) holds, then all the results of Chap. 3 are true. In particular, Theorem 3.2, a turnpike result, is true. The next result shows that if the turnpike property established by Theorem 3.2 holds, then (B1) holds too.

Theorem 4.1 *Assume that for each $\epsilon > 0$ there exist a natural number τ and $\delta > 0$ such that the following property holds:*

(i) for each integer $T > 2\tau$ and each program $\{z_t\}_{t=0}^{T}$ satisfying

$$\sum_{t=0}^{T-1} u(z_t, z_{t+1}) \ge U(z_0, z_T, 0, T) - \delta$$

the inequality

$$\rho(z_t, \bar{x}) \le \epsilon$$

holds for all integers $t = \tau, \ldots, T - \tau$.

Then for each good program $\{x_t\}_{t=0}^{\infty}$,

$$\lim_{t \to \infty} x_t = \bar{x}.$$

Proof Assume that $\{x_t\}_{t=0}^{\infty}$ is a good program. Then there exists $M > 0$ such that for each natural number T,

$$\left| \sum_{t=0}^{T-1} u(x_t, x_{t+1}) - T\mu \right| \leq M. \tag{4.3}$$

Let $\epsilon > 0$ and $\delta > 0$ and a natural number τ be such that property (i) holds. We show that there exists an integer $T_\delta \geq 0$ such that for each integer $T > T_\delta$,

$$\sum_{t=T_\delta}^{T-1} u(x_t, x_{t+1}) \geq U(x_{T_\delta}, x_T, T_\delta, T) - \delta. \tag{4.4}$$

Assume the contrary. Then there exists a strictly increasing sequence of integers $\{T_i\}_{i=0}^{\infty}$ such that $T_0 = 0$ and for each integer $i \geq 1$,

$$\sum_{t=T_i}^{T_{i+1}-1} u(x_t, x_{t+1}) < U(x_{T_i}, x_{T_{i+1}}, T_i, T_{i+1}) - \delta. \tag{4.5}$$

By (4.5), there exists a program $\{y_t\}_{t=0}^{\infty}$ such that for each integer $i \geq 0$,

$$y_{T_i} = x_{T_i},$$

$$\sum_{t=T_i}^{T_{i+1}-1} u(x_t, x_{t+1}) < \sum_{t=T_i}^{T_{i+1}-1} u(y_t, y_{t+1}) - \delta.$$

In view of the relation above and (4.3) for each natural number q,

$$\sum_{t=0}^{T_q-1} u(y_t, y_{t+1}) - T_q\mu$$

$$= \sum_{t=0}^{T_q-1} u(y_t, y_{t+1}) - \sum_{t=0}^{T_q-1} u(x_t, x_{t+1}) + \sum_{t=0}^{T_q-1} u(x_t, x_{t+1}) - T_q\mu$$

$$\geq q\delta - M \to \infty \text{ as } q \to \infty.$$

This contradicts Proposition 3.1. The contradiction we have reached proves that there exists an integer $T_\delta \geq 0$ such that the following property holds:

(ii) for each integer $T > T_\delta$, (4.4) is true.

Properties (i) and (ii), (4.9), and the choice of δ, τ imply that for each integer $T > T_\delta + 2\tau$, we have

$$\rho(x_t, \bar{x}) \le \epsilon, \ t = \tau + T_\delta, \dots, T - \tau.$$

This implies that

$$\rho(x_t, \bar{x}) \le \epsilon \text{ for all integers } t \ge \tau + T_\delta.$$

Theorem 4.1 is proved.

4.3 Generic Results

We use the notation, definitions, and assumptions introduced in Sect. 4.1. Denote by $C(\Omega)$ the set of all continuous functions on Ω. Clearly, $(C(\Omega), \|\cdot\|)$ is a Banach space. For each $u_1, u_2 \in C(\Omega)$ set

$$d(u_1, u_2) = \|u_1 - u_2\|.$$

Let $v \in C(\Omega)$. By Theorem 2.9 there exist

$$\mu(v) = \lim_{p \to \infty} \widehat{U}(v, 0, p) p^{-1} \tag{4.6}$$

and $M_v > 0$ such that

$$\left| \mu(v) - \widehat{U}(v, 0, p) p^{-1} \right| \le M_v / p \tag{4.7}$$

for all natural numbers p.

Assume that there exists $z \in K$ such that

$$(z, z) \in \Omega.$$

Denote by $\mathcal{M}$ the set of all $v \in C(\Omega)$ such that

$$\mu(v) = \sup\{v(x, x) : \ x \in K \text{ and } (x, x) \in \Omega\}. \tag{4.8}$$

It is not difficult to see that $\mathcal{M}$ is a closed subset of $C(\Omega)$. We consider the complete metric space $(\mathcal{M}, d)$. Suppose that the following assumption holds.

(B4) There exists a natural number L_* such that for each $\epsilon > 0$ there exists $\delta > 0$ such that for each $y_1, y_2 \in K$ satisfying $\rho(y_1, y_2) \le \delta$ there is a program $\{x_t\}_{t=0}^{L_*}$ such that

$$x_0 = y_1, \ x_{L_*} = y_2$$

and

$$\rho(x_t, \bar{x}) \leq \epsilon, \ t = 0, \ldots, L_*.$$

Clearly, (B4) is a strong version of (B3).

Let $v \in \mathcal{M}$. Recall that a program $\{x_t\}_{t=0}^{\infty}$ is called (v)-good if the sequence

$$\left\{ \sum_{t=0}^{T-1} \{v(x_t, x_{t+1}) - \mu(v)\} \right\}_{T=1}^{\infty}$$

is bounded.

Theorem 4.2 *There exists a set $\mathcal{F} \subset \mathcal{M}$ which is a countable intersection of open everywhere dense subsets of $\mathcal{M}$ such that for each $v \in \mathcal{F}$ the following properties hold:*

1. there exists a unique point $x_v \in K$ such that

$$\mu(v) = v(x_v, x_v).$$

2. for each (v)-good program $\{x_t\}_{t=0}^{\infty}$,

$$\lim_{t \to \infty} \rho(x_t, \bar{x}) = 0.$$

It is clear that if $u \in \mathcal{F}$, then all the results of Chap. 3 hold for u.

Proof of Theorem 4.2 Let $v \in \mathcal{M}$ and $r > 0$. There exists $x_v \in K$ such that

$$(x_v, x_v) \in \Omega, \ \mu(v) = v(x_v, x_v). \tag{4.9}$$

For $(x, y) \in \Omega$ define

$$v_r(x, y) = v(x, y) - r\rho(x, x_v). \tag{4.10}$$

It is not difficult to see that $v_r \in \mathcal{M}$ and

$$\mu(v_r) = \mu(v) = v(x_v, x_v). \tag{4.11}$$

Clearly, the set

$$\{v_r : \ v \in \mathcal{M}, \ r \in (0, 1]\}$$

is an everywhere dense subset of $\mathcal{M}$.

Lemma 4.3 *Let $v \in \mathcal{M}$ and $r \in (0, 1]$. Assume that $\{x_t\}_{t=0}^{\infty}$ is a (v_r)-good program. Then*

$$\lim_{t \to \infty} \rho(x_t, x_v) = 0.$$

Proof There exists $M > 0$ such that

$$\left| \sum_{t=0}^{T-1} v_r(x_t, x_{t+1}) - Tv(x_v, x_v) \right| \leq M \text{ for all natural numbers T.} \qquad (4.12)$$

Proposition 3.1, (4.10), and (4.12) imply that for all natural numbers T,

$$- M \leq \sum_{t=0}^{T-1} v(x_t, x_{t+1}) - Tv(x_v, x_v) - r \sum_{t=0}^{T-1} \rho(x_t, x_v). \qquad (4.13)$$

In view of (4.13), the program $\{x_t\}_{t=0}^{\infty}$ is (v)-good and

$$\lim_{T \to \infty} \sum_{t=0}^{T-1} \rho(x_t, x_v) < \infty.$$

This implies that

$$\lim_{t \to \infty} \rho(x_t, x_v).$$

Lemma 4.3 is proved.

In view of Lemma 4.3, all the results of Chap. 3 are true for $u = v_r$.

Completion of the Proof of Theorem 4.2 Let $v \in \mathcal{M}$, $r \in (0, 1]$ and n be a natural number. Theorem 3.12 implies that there exist an open neighborhood $\mathcal{U}(v, r, n)$ of v_r in the metric space $(\mathcal{M}, d)$, a natural number $\tau(v, r, n)$, and $\gamma(v, r, n) > 0$ such that the following property holds:

(i) for each integer $T > 2\tau(v, r, n)$, each $u_t \in \mathcal{U}(v, r, n)$, $t = 0, \ldots, T - 1$ and each program $\{x_t\}_{t=0}^{T}$ which satisfies

$$\sum_{t=0}^{T-1} u_t(x_t, x_{t+1}) \geq U\left(\{u_t\}_{t=0}^{T-1} x_0, x_T\right) - \gamma(v, r, n)$$

we have

$$\rho(x_t, x_v) \leq 1/n, \ t = \tau(v, r, n), \ldots, T - \tau(v, r, n).$$

Define

$$\mathcal{F} = \cap_{p=1}^{\infty} \cup \{\mathcal{U}(v, r, n) : \ v \in \mathcal{M}, \ r \in (0, 1], \ n \geq p \text{ is an integer}\}. \qquad (4.14)$$

Clearly, $\mathcal{F}$ is a countable intersection of open everywhere dense sets in $(\mathcal{M}, d)$.

Let

$$w \in \mathcal{F}, \ \epsilon \in (0, 1). \tag{4.15}$$

Choose a natural number p such that

$$p > 4\epsilon^{-1}. \tag{4.16}$$

It follows from (4.14) and (4.15) that there exist $v \in \mathcal{M}$, $r \in (0, 1]$, and an integer $n \geq p$ such that

$$w \in \mathcal{U}(v, r, n). \tag{4.17}$$

Assume that

$$\xi_1, \xi_2 \in K,$$

$$w(\xi_i, \xi_i) = \mu(w), \ i = 1, 2. \tag{4.18}$$

It is not difficult to see that for each $i = 1, 2$ and each integer $T > 0$,

$$U(w, \xi_i, \xi_i, 0, T) = T w(\xi_i, \xi_i).$$

Combined with property (i) this implies that

$$\rho(x_v, \xi_i) \leq n^{-1}, \ i = 1, 2. \tag{4.19}$$

By (4.16) and (4.19),

$$\rho(\xi_1, \xi_2) \leq 2/n \leq 2/p < \epsilon.$$

Since ϵ is an arbitrary positive number we conclude that $\xi_1 = \xi_2$ and

$$\{\xi \in K : \ (\xi, \xi) \in \Omega, \ w(\xi, , \xi) = \mu(w)\} = \{\xi_1\},$$

a singleton.

Assume that an integer $T > 2\tau(v, r, n)$ and that a program $\{x_t\}_{t=0}^{T}$ satisfies

$$\sum_{t=0}^{T-1} w(x_t, x_{t+1}) \geq U(w, x_0, x_T, 0, T) - \gamma(v, r, n).$$

Together with property (i), (4.16), (4.17), and (4.19) this implies that for all integers
$t = \tau(v, r, n), \ldots, T - \tau(v, r, n)$,

$$\rho(x_t, x_v) \leq 1/n$$

and

$$\rho(x_t, \xi_1) \leq \rho(x_t, x_v) + \rho(x_v, \xi_1) \leq 2/n \leq 2/p < \epsilon.$$

Combined with Theorem 4.1 this implies that every (w)-good program converges
to ξ_1. Theorem 4.2 is proved.

Chapter 5
Structure of Solutions in the Regions Close to the Endpoints

In this chapter we continue to study the structure of approximate solutions of the autonomous nonconcave discrete-time optimal control system with a compact metric space of states. This control system is described by a bounded upper semicontinuous objective function which determines an optimality criterion. In the turnpike theory, it is known that approximate solutions are determined mainly by the objective function, and are essentially independent of the choice of time intervals and data, except in regions close to the endpoints of the time intervals. In this chapter our main goal is to analyze the structure of approximate solutions in regions close to the endpoints of the time intervals.

5.1 Preliminaries

We continued to study the class of optimal control problems studied in Chap. 3 (see Sects. 3.1 and 3.8) using the same notation, definitions, and assumptions.

Let (K, ρ) be a compact metric space and let Ω be a nonempty closed subset of $K \times K$.

Let $u : \Omega \to R^1$ be a bounded upper semicontinuous function.

In this chapter we suppose that the following assumption, introduced in Sect. 2.2, holds.

(A)   There exists a natural number $\bar{L}$ such that for each $y, z \in K$ there is a (Ω)-program $\{x_t\}_{t=0}^{\bar{L}}$ such that $x_0 = y$ and $x_{\bar{L}} = z$.

We also assume that there is $\bar{x} \in K$ such that

$$(\bar{x}, \bar{x}) \in \Omega \text{ and } u(\bar{x}, \bar{x}) = \mu(u) = \mu. \tag{5.1}$$

© The Author(s), under exclusive license to Springer Nature Switzerland AG 2019
A. J. Zaslavski, *Optimal Control Problems Arising in Forest Management*,
SpringerBriefs in Optimization, https://doi.org/10.1007/978-3-030-23587-1_5

We also assume that the following assumptions (B1)–(B3) introduced in Sect. 3.1 hold.

(B1) For each (u, Ω)-good program $\{x_t\}_{t=0}^{\infty}$, $\lim_{t\to\infty} x_t = \bar{x}$.

(B2) u is continuous at $(\bar{x}, \bar{x})$ with respect to Ω.

(B3) There exists a natural number L_* such that for each $\epsilon > 0$ there exists $\delta > 0$ such that for each $y_1, y_2 \in K$ satisfying $\rho(y_1, \bar{x})$, $\rho(\bar{x}, y_2) \leq \delta$ there is an (Ω)-program $\{x_t\}_{t=0}^{L_*}$ such that $x_0 = y_1$, $x_{L_*} = y_2$ and $\rho(x_t, \bar{x}) \leq \epsilon$, $t = 0, \ldots, L_*$.

We study the structure of approximate solutions of problems (P_1)–(P_3) (see Sect. 3.1).

We define a function $\pi^u(x)$, $x \in K$ which plays an important role in our study. Let $x \in K$. Set

$$\pi^u(x) = \sup \left\{ \limsup_{T\to\infty} \sum_{t=0}^{T-1} (u(x_t, x_{t+1}) - v(\bar{x}, \bar{x})) : \right.$$

$$\left. \{x_t\}_{t=0}^{\infty} \text{ is an } (\Omega) - \text{program such that } x_0 = x \right\}. \tag{5.2}$$

In view of Theorems 2.6 and 2.9, there exists $M_u > 0$ such that

$$- M_u \leq \pi^u(x) \leq M_u \text{ for all } x \in K. \tag{5.3}$$

Let $x \in K$. By (5.3),

$$\pi^u(x) = \sup \left\{ \limsup_{T\to\infty} \sum_{t=0}^{T-1} (u(x_t, x_{t+1}) - u(\bar{x}, \bar{x})) : \right.$$

$$\left. \{x_t\}_{t=0}^{\infty} \text{ is an } (u, \Omega) - \text{good program such that } x_0 = x \right\}. \tag{5.4}$$

Denote by $\mathcal{P}(u, x)$ the set of all (u, Ω)-overtaking optimal programs $\{x_t\}_{t=0}^{\infty}$ satisfying $x_0 = x$. By Theorem 3.4, the set $\mathcal{P}(u, x)$ is nonempty. Definition (5.2) implies the following result.

Proposition 5.1 *Let $T \geq 1$ be an integer and $\{x_t\}_{t=0}^{T}$ be a (Ω)-program. Then for each integer $t = 0, \ldots, T - 1$,*

$$\pi^u(x_t) \geq u(x_t, x_{t+1}) - u(\bar{x}, \bar{x}) + \pi^u(x_{t+1}).$$

The next result follows from the definition of (u, Ω)-overtaking optimal programs.

Proposition 5.2 *Let $\{x_t\}_{t=0}^{\infty}$ be a (u, Ω)-overtaking optimal program. Then*

$$\pi^u(x_0) = \limsup_{T \to \infty} \sum_{t=0}^{T-1} (u(x_t, x_{t+1}) - u(\bar{x}, \bar{x})).$$

Corollary 5.3 *Let $\{x_t\}_{t=0}^{\infty}$ be a (u, Ω)-overtaking optimal program. Then for any integer $t \geq 0$,*

$$\pi^u(x_t) = u(x_t, x_{t+1}) - u(\bar{x}, \bar{x}) + \pi^u(x_{t+1}). \tag{5.5}$$

Set

$$\sup(\pi^u) = \sup\{\pi^u(z) : z \in K\}. \tag{5.6}$$

Proposition 5.4 $\pi^u(\bar{x}) = 0$.

Proof Set $x_t = \bar{x}$ for all integers $t \geq 0$. By Theorem 3.5 and (5.1), the program $\{x_t\}_{t=0}^{\infty}$ is a (u, Ω)-overtaking optimal. In view of Proposition 5.2, $\pi^u(\bar{x}) = 0$. Proposition 5.4 is proved.

Proposition 5.5 *The function π^u is continuous at $\bar{x}$.*

Proof Let $\epsilon > 0$. In view of (B2), there exists $\epsilon_1 \in (0, \epsilon)$ such that

$$|u(x, y) - u(\bar{x}, \bar{x})| \leq \epsilon(4L_*)^{-1}$$

for each $(x, y) \in \Omega$ satisfying $\rho(x, \bar{x})$, $\rho(y, \bar{x}) \leq \epsilon_1$. $\tag{5.7}$

By (B3), there exists $\delta \in (0, \epsilon_1)$ such that the following property holds:

(i) for each $z_1, z_2 \in K$ satisfying

$$\rho(z_i, \bar{x}) \leq \delta, \ i = 1, 2$$

there exists an (Ω)-program $\{x_t\}_{t=0}^{L_*}$ such that

$$x_0 = z_1, \ x_{L_*} = z_2$$

and

$$\rho(x_t, \bar{x}) \leq \epsilon_1, \ t = 0, \ldots, L_*.$$

Assume that

$$z_1, z_2 \in K, \ \rho(z_i, \bar{x}) \leq \delta, \ i = 1, 2. \tag{5.8}$$

Property (i) and (5.8) imply that there exists an (Ω)-program $\{x_t\}_{t=0}^{L_*}$ such that

$$x_0 = z_1, \quad x_{L_*} = z_2, \tag{5.9}$$

and

$$\rho(x_t, \bar{x}) \le \epsilon_1, \quad t = 0, \ldots, L_*. \tag{5.10}$$

By (5.7) and (5.10), for all $t = 0, \ldots, L_* - 1$,

$$|u(x_t, x_{t+1}) - u(\bar{x}, \bar{x})| \le \epsilon(4L_*)^{-1}.$$

It follows from the relation above, (5.2), and (5.9) that

$$\pi^u(z_1) = \pi^u(x_0)$$

$$\ge \sum_{t=0}^{L_*-1} (u(x_t, x_{t+1}) - u(\bar{x}, \bar{x})) + \pi^u(x_{L_*})$$

$$\ge -4^{-1}(\epsilon/L_*)L_* + \pi^u(z_2) = \pi^u(z_2) - \epsilon.$$

Proposition 5.5 is proved.

Proposition 5.6 *Assume that* $x_0 \in K$ *and* $\{x_t\}_{t=0}^{\infty} \in \mathcal{P}(u, x_0)$. *Then*

$$\pi^u(x_0) = \lim_{T \to \infty} \sum_{t=0}^{T-1} (u(x_t, x_{t+1}) - u(\bar{x}, \bar{x})).$$

Proof It follows from (B1), Propositions 5.4, 5.5, and Corollary 5.3 that

$$\pi^u(x_0) = \lim_{T \to \infty} (\pi^u(x_0) - \pi^u(x_T))$$

$$= \lim_{T \to \infty} \sum_{t=0}^{T-1} (u(x_t, x_{t+1}) - u(\bar{x}, \bar{x})).$$

Proposition 5.6 is proved.

Proposition 5.7 *The function* $\pi^u : K \to R^1$ *is upper semicontinuous.*

Proof Assume that $\{x^{(i)}\}_{i=1}^{\infty} \subset K$, $x \in K$ and

$$\lim_{i \to \infty} x^{(i)} = x. \tag{5.11}$$

We show that

$$\pi^u(x) \geq \limsup_{i \to \infty} \pi^u \left(x^{(i)} \right).$$

We may assume without loss of generality that

$$\limsup_{i \to \infty} \pi^u \left(x^{(i)} \right) = \lim_{i \to \infty} \pi^u \left(x^{(i)} \right). \tag{5.12}$$

By Theorem 3.4 and Proposition 5.6, for each integer $i \geq 1$, there exists a (u, Ω)-overtaking optimal program $\{x_t^{(i)}\}_{t=0}^{\infty}$ such that

$$x_0^{(i)} = x^{(i)}, \tag{5.13}$$

$$\pi^u \left(x^{(i)} \right) = \lim_{T \to \infty} \sum_{t=0}^{T-1} \left(u(x_t^{(i)}, x_{t+1}^{(i)}) - u(\bar{x}, \bar{x}) \right). \tag{5.14}$$

Extracting a subsequence and re-indexing, if necessary, we may assume without loss of generality that for each integer $t \geq 0$ there exists

$$x_t = \lim_{i \to \infty} x_t^{(i)}. \tag{5.15}$$

In view of (5.15), $\{x_t\}_{t=0}^{\infty}$ is an (Ω)-program.

Let $\epsilon > 0$. By Propositions 5.4 and 5.5, there exists $\delta \in (0, \epsilon)$ such that for each $x \in K$ satisfying $\rho(x, \bar{x}) \leq \delta$,

$$|\pi^u(x)| \leq \epsilon/2. \tag{5.16}$$

Since $\{x_t^{(i)}\}_{t=0}^{\infty}$, $i = 1, 2, \ldots$ are (u, Ω)-overtaking optimal programs it follows from Theorem 3.2 that there exists a natural number L_1 such that for all sufficiently large natural numbers i,

$$\rho \left(x_t^{(i)}, \bar{x} \right) \leq \delta \text{ for all integers } t \geq L_1. \tag{5.17}$$

Let $T \geq L_1$ be an integer. By Corollary 5.3, (5.16), and (5.17), for all integers $i \geq 1$,

$$\sum_{t=0}^{T-1} (u(x_t^{(i)}, x_{t+1}^{(i)}) - u(\bar{x}, \bar{x}))$$

$$= \pi^u(x_0^{(i)}) - \pi^u(x_T^{(i)}) \geq \pi^u(x_0^{(i)}) - \epsilon/2.$$

In view of the relation above and the upper semicontinuity of u, for all integers $T \geq L_1$,

$$\sum_{t=0}^{T-1}(u(x_t, x_{t+1}) - u(\bar{x}, \bar{x}))$$

$$\geq \limsup_{i \to \infty} \sum_{t=0}^{T-1}\left(u(x_t^{(i)}, x_{t+1}^{(i)}) - u(\bar{x}, \bar{x})\right)$$

$$\geq \limsup_{i \to \infty} \pi^u\left(x_0^{(i)}\right) - \epsilon/2.$$

By (5.2), (5.11), (5.13), (5.15), and the relation above,

$$\pi^u(x) \geq \limsup_{T \to \infty} \sum_{t=0}^{T-1}(u(x_t, x_{t+1}) - u(\bar{x}, \bar{x})) \geq \limsup_{i \to \infty} \pi^u\left(x^{(i)}\right) - \epsilon/2.$$

Since ϵ is any positive number we conclude that

$$\pi^u(x) \geq \limsup_{i \to \infty} \pi^u\left(x^{(i)}\right).$$

Proposition 5.7 is proved.

Proposition 5.8 *Let $\{x_t\}_{t=0}^{\infty}$ be an (Ω)-program such that for all integers $t \geq 0$,*

$$u(x_t, x_{t+1}) - u(\bar{x}, \bar{x}) = \pi^u(x_t) - \pi^u(x_{t+1}). \tag{5.18}$$

Then $\{x_t\}_{t=0}^{\infty}$ is a (u, Ω)-overtaking optimal program.

Proof Proposition 5.1 and (5.18) imply that $\{x_t\}_{t=0}^{\infty}$ is a (u, Ω)-locally optimal program. By Theorem 3.5, $\{x_t\}_{t=0}^{\infty}$ is a (u, Ω)-overtaking optimal program. Proposition 5.8 is proved.

5.2   Lagrange Problems

In order to study the structure of solutions of the problems in the regions close to the end points we introduce the following notation and definitions.

Set

$$\bar{\Omega} = \{(x, y) \in K \times K : (y, x) \in \Omega\}. \tag{5.19}$$

Clearly, $\bar{\Omega}$ is a nonempty closed subset of $K \times K$.

For each function $w : \Omega \to R^1$ define $\bar{w} : \bar{\Omega} \to R^1$ by

$$\bar{w}(x, y) = w(y, x), \ (x, y) \in \bar{\Omega}. \tag{5.20}$$

Let $0 \le T_1 < T_2$ be integers and let $\{x_t\}_{t=T_1}^{T_2}$ be an (Ω)-program. Define $\{\bar{x}_t\}_{t=T_1}^{T_2} \subset K$ by

$$\bar{x}_t = x_{T_2-t+T_1}, \ t = T_1, \dots, T_2. \tag{5.21}$$

Clearly, $\{\bar{x}_t\}_{t=T_1}^{T_2}$ is an $(\bar{\Omega})$-program.

Assume that $u_t : \Omega \to R^1$, $t = T_1, \dots, T_2 - 1$. It is easy to see that

$$\sum_{t=T_1}^{T_2-1} \bar{u}_{T_2-t+T_1-1}(\bar{x}_t, \bar{x}_{t+1})$$

$$= \sum_{t=T_1}^{T_2-1} u_{T_2-t+T_1-1}(x_{T_2-t+T_1-1}, x_{T_2-t+T_1})$$

$$= \sum_{t=T_1}^{T_2-1} u_t(x_t, x_{t+1}). \tag{5.22}$$

The next result easily follows from (5.22).

Proposition 5.9 *Let $T \ge 1$ be an integer, $M \ge 0$, $u_t : \Omega \to R^1$, $t = 0, \dots, T$ and $\{x_t^{(i)}\}_{t=0}^T$, $i = 1, 2$ are (Ω)-programs. Then*

$$\sum_{t=0}^{T-1} u_t\left(x_t^{(1)}, x_{t+1}^{(1)}\right) \ge \sum_{t=0}^{T-1} u_t\left(x_t^{(2)}, x_{t+1}^{(2)}\right) - M$$

if and only if

$$\sum_{t=0}^{T-1} \bar{u}_{T-t-1}\left(\bar{x}_t^{(1)}, \bar{x}_{t+1}^{(1)}\right) \ge \sum_{t=0}^{T-1} \bar{u}_{T-t-1}\left(x_t^{(2)}, x_{t+1}^{(2)}\right) - M.$$

Let T_1, T_2 be integers such that $T_1 < T_2$ and $u_t : \Omega \to R^1, t = T_1, \dots T_2 - 1$ be bounded functions. For each $z_0, z_1 \in K$ set

$$U\left(\{u_t\}_{t=T_1}^{T_2-1}, \Omega\right) = \sup \left\{ \sum_{t=T_1}^{T_2-1} u_t(x_t, x_{t+1}) : \{x_t\}_{t=T_1}^{T_2} \text{ is an } (\Omega)\text{-program} \right\},$$

$$\tag{5.23}$$

$$U\left(\{u_t\}_{t=T_1}^{T_2-1}, \Omega, z_0\right) = \sup \left\{ \sum_{t=T_1}^{T_2-1} u_t(x_t, x_{t+1}) : \{x_t\}_{t=T_1}^{T_2} \text{ is an } (\Omega)\text{-program}\right.$$

$$\left. \text{such that } x_{T_1} = z_0 \right\}, \tag{5.24}$$

$$\widehat{U}\left(\{u_t\}_{t=T_1}^{T_2-1}, \Omega, z_1\right) = \sup \left\{ \sum_{t=T_1}^{T_2-1} u_t(x_t, x_{t+1}) : \{x_t\}_{t=T_1}^{T_2} \text{ is an } (\Omega)\text{-program}\right.$$

$$\left. \text{such that } x_{T_2} = z_1 \right\}, \tag{5.25}$$

$$U\left(\{u_t\}_{t=T_1}^{T_2-1}, \Omega, z_0, z_1\right) = \sup \left\{ \sum_{t=T_1}^{T_2-1} u_t(x_t, x_{t+1}) : \{x_t\}_{t=T_1}^{T_2} \text{ is an } (\Omega)\text{-program}\right.$$

$$\left. \text{such that } x_{T_1} = z_0, \ x_{T_2} = z_1 \right\}. \tag{5.26}$$

Let T_1, T_2 be integers such that $T_1 < T_2$ and $w_t : \bar{\Omega} \to R^1, t = T_1, \ldots T_2 - 1$ be bounded functions. For each $z_0, z_1 \in K$ set

$$U\left(\{w_t\}_{t=T_1}^{T_2-1}, \bar{\Omega}, z_0\right) = \sup \left\{ \sum_{t=T_1}^{T_2-1} w_t(x_t, x_{t+1}) : \{x_t\}_{t=T_1}^{T_2} \text{ is an } (\bar{\Omega})\text{-program}\right.$$

$$\left. \text{such that } x_{T_1} = z_0 \right\}, \tag{5.27}$$

$$U\left(\{w_t\}_{t=T_1}^{T_2-1}, \bar{\Omega}, z_0, z_1\right) = \sup \left\{ \sum_{t=T_1}^{T_2-1} w_t(x_t, x_{t+1}) : \{x_t\}_{t=T_1}^{T_2} \text{ is an } (\bar{\Omega})\text{-program}\right.$$

$$\left. \text{such that } x_{T_1} = z_0, \ x_{T_2} = z_1 \right\}, \tag{5.28}$$

$$\widehat{U}\left(\{w_t\}_{t=T_1}^{T_2-1}, \bar{\Omega}, z_1\right) = \sup\left\{\sum_{t=T_1}^{T_2-1} w_t(x_t, x_{t+1}): \{x_t\}_{t=T_1}^{T_2} \text{ is an } (\bar{\Omega})\text{-program}\right.$$

such that $x_{T_2} = z_1\}$, (5.29)

$$U\left(\{w_t\}_{t=T_1}^{T_2-1}, \bar{\Omega}\right) = \sup\left\{\sum_{t=T_1}^{T_2-1} w_t(x_t, x_{t+1}): \{x_t\}_{t=T_1}^{T_2} \text{ is an } (\bar{\Omega})\text{-program}\right\}.$$

(5.30)

Proposition 5.9 implies the following result.

Proposition 5.10 *Let $T \geq 1$ be an integer, $M \geq 0$, $u_t : \Omega \to R^1$, $t = 0, \ldots T - 1$ be bounded functions and $\{x_t\}_{t=0}^{T}$ be an (Ω)-program. Then $\{\bar{x}_t\}_{t=0}^{T}$ is an $(\bar{\Omega})$-program and the following assertions hold:*
if $\sum_{t=0}^{T-1} u_t(x_t, x_{t+1}) \geq U\left(\{u_t\}_{t=0}^{T-1}, \Omega\right) - M$, then

$$\sum_{t=0}^{T-1} \bar{u}_{T-t-1}(\bar{x}_t, \bar{x}_{t+1}) \geq U\left(\{\bar{u}_{T-t-1}\}_{t=0}^{T-1}, \bar{\Omega}\right) - M;$$

if $\sum_{t=0}^{T-1} u_t(x_t, x_{t+1}) \geq U\left(\{u_t\}_{t=0}^{T-1}, \Omega, x_0, x_T\right) - M$, then

$$\sum_{t=0}^{T-1} \bar{u}_{T-t-1}(\bar{x}_t, \bar{x}_{t+1}) \geq U\left(\{\bar{u}_{T-t-1}\}_{t=0}^{T-1}, \bar{\Omega}, \bar{x}_0, \bar{x}_T\right) - M;$$

if $\sum_{t=0}^{T-1} u_t(x_t, x_{t+1}) \geq U\left(\{u_t\}_{t=0}^{T-1}, \Omega, x_0\right) - M$, then

$$\sum_{t=0}^{T-1} \bar{u}_{T-t-1}(\bar{x}_t, \bar{x}_{t+1}) \geq \widehat{U}\left(\{\bar{u}_{T-t-1}\}_{t=0}^{T-1}, \bar{\Omega}, \bar{x}_T\right) - M;$$

if $\sum_{t=0}^{T-1} u_t(x_t, x_{t+1}) \geq \widehat{U}\left(\{u_t\}_{t=0}^{T-1}, \Omega, x_T\right) - M$, then

$$\sum_{t=0}^{T-1} \bar{u}_{T-t-1}(\bar{x}_t, \bar{x}_{t+1}) \geq U\left(\{\bar{u}_{T-t-1}\}_{t=0}^{T-1}, \bar{\Omega}, \bar{x}_0\right) - M.$$

Note that the pair $(\bar{u}, \bar{\Omega})$ satisfies assumptions (A), (B2), and (B3), and that

$$\mu(\bar{u}) = \mu(u) = \mu.$$

It follows from Theorems 3.2 and 4.1 and Proposition 5.10 that (B1) also holds for the pair $(\bar{u}, \bar{\Omega})$. Therefore all the results presented above for the pair (u, Ω) are also true for the pair $(\bar{u}, \bar{\Omega})$.

We prove the following results which describe the structure of approximate solutions of our optimal control problems in the regions close to the endpoints.

Theorem 5.11 *Let* $\tau_0 \geq 1$ *be an integer and* $\epsilon > 0$. *Then there exist* $\delta > 0$ *and an integer* $T_0 \geq \tau_0$ *such that for each integer* $T \geq T_0$, *each finite sequence of bounded functions* $u_t : \Omega \to R^1$, $t = 0, \ldots T - 1$ *satisfying*

$$\|u_t - u\| \leq \delta, \ t = 0 \ldots, T - 1$$

and each (Ω)-*program* $\{x_t\}_{t=0}^{T}$ *which satisfies*

$$\sum_{t=0}^{T-1} u_t(x_t, x_{t+1}) \geq U\left(\{u_t\}_{t=0}^{T-1}, \Omega, x_0\right) - \delta$$

there exists a $(\bar{u}, \bar{\Omega})$-*overtaking optimal program* $\{x_t^*\}_{t=0}^{\infty}$

$$\pi^{\bar{u}}(x_0^*) = \sup(\pi^{\bar{u}}),$$

$$\rho(x_{T-t}, x_t^*) \leq \epsilon, \ t = 0, \ldots, \tau_0.$$

Theorem 5.12 *Let* $\tau_0 \geq 1$ *be an integer and* $\epsilon > 0$. *Then there exist* $\delta > 0$ *and an integer* $T_0 \geq \tau_0$ *such that for each integer* $T \geq T_0$, *each finite sequence of bounded functions* $u_t : \Omega \to R^1$, $t = 0, \ldots T - 1$ *satisfying*

$$\|u_t - u\| \leq \delta, \ t = 0 \ldots, T - 1$$

and each (Ω)-*program* $\{x_t\}_{t=0}^{T}$ *which satisfies*

$$\sum_{t=0}^{T-1} u_t(x_t, x_{t+1}) \geq U\left(\{u_t\}_{t=0}^{T-1}, \Omega\right) - \delta$$

there exist a (u, Ω)-*overtaking optimal program* $\{y_t^{(1)}\}_{t=0}^{\infty}$ *and a* $(\bar{u}, \bar{\Omega})$-*overtaking optimal program* $\{y_t^{(2)}\}_{t=0}^{\infty}$ *such that*

$$\pi^u\left(y_0^{(1)}\right) = \sup(\pi^u), \ \pi^{\bar{u}}\left(y_0^{(2)}\right) = \sup(\pi^{\bar{u}})$$

and that for all integers $t = 0, \ldots, \tau_0$,

$$\rho\left(x_{T-t}, y_t^{(2)}\right) \leq \epsilon, \ \rho\left(x_t, y_t^{(1)}\right) \leq \epsilon.$$

Theorem 5.13 *Let $\tau_0 \geq 1$ be an integer and $\epsilon > 0$. Then there exist $\delta > 0$ and an integer $T_0 \geq \tau_0$ such that for each integer $T \geq T_0$, each finite sequence of bounded functions $u_t : \Omega \to R^1$, $t = 0, \ldots T - 1$ satisfying*

$$\|u_t - u\| \leq \delta, \ t = 0 \ldots, T - 1$$

and each (Ω)-program $\{x_t\}_{t=0}^{T}$ which satisfies

$$\sum_{t=0}^{T-1} u_t(x_t, x_{t+1}) \geq \widehat{U}\left(\{u_t\}_{t=0}^{T-1}, \Omega, x_T\right) - \delta$$

there exists a (u, Ω)-overtaking optimal program $\{x_t^*\}_{t=0}^{\infty}$ such that

$$\pi^u(x_0^*) = \sup(\pi^u),$$
$$\rho(x_t, x_t^*) \leq \epsilon, \ t = 0, \ldots, \tau_0.$$

Theorem 5.13 is proved in Sect. 5.4 while Sect. 5.3 contains an auxiliary result. Theorem 5.11 follows from Theorem 5.13 applied with $(\bar{u}, \bar{\Omega})$ and Proposition 5.10. Theorem 5.12 easily follows from Theorems 5.11 and 5.13.

5.3   An Auxiliary Result for Theorem 5.13

Lemma 5.14 *Let $T_0 \geq 1$ be an integer and $\epsilon \in (0, 1)$. Then there exists $\delta \in (0, \epsilon)$ such that for each (Ω)-program $\{x_t\}_{t=0}^{T_0}$ which satisfies*

$$\pi^u(x_0) \geq \sup(\pi^u) - \delta,$$

$$\sum_{t=0}^{T_0-1} (u(x_t, x_{t+1}) - u(\bar{x}, \bar{x})) - \pi^u(x_0) + \pi^u(x_T) \geq -\delta,$$

there exists a (u, Ω)-overtaking optimal program $\{z_t\}_{t=0}^{\infty}$ such that

$$\pi^u(z_0) = \sup(\pi^u),$$
$$\rho(z_t, x_t) \leq \epsilon, \ t = 0, \ldots, T_0. \tag{5.31}$$

Proof Assume that the lemma does not hold. Then there exist a sequence $\{\delta_k\}_{k=1}^{\infty} \subset (0, 1)$ and a sequence of (Ω)-programs $\{x_t^{(k)}\}_{t=0}^{T_0}$, $k = 1, 2, \ldots$ such that

$$\lim_{k \to \infty} \delta_k = 0 \tag{5.32}$$

and that for each integer $k \geq 1$ and each (u, Ω)-overtaking optimal program $\{z_t\}_{t=0}^{\infty}$ satisfying (5.31),

$$\pi^u\left(x_0^{(k)}\right) \geq \sup(\pi^u) - \delta_k, \tag{5.33}$$

$$\sum_{t=0}^{T_0-1} \left(u\left(x_t^{(k)}, x_{t+1}^{(k)}\right) - u(\bar{x}, \bar{x})\right) - \pi^u\left(x_0^{(k)}\right) + \pi^u\left(x_{T_0}^{(k)}\right) \geq -\delta_k \tag{5.34}$$

we have

$$\max\left\{\rho\left(z_t, x_t^{(k)}\right) : t = 0, \ldots, T_0\right\} > \epsilon. \tag{5.35}$$

Extracting a subsequence and re-indexing, if necessary, we may assume without loss of generality that for each integer $t \in [0, T_0]$ there exists

$$x_t = \lim_{k \to \infty} x_t^{(k)}. \tag{5.36}$$

Proposition 5.7, (5.32), and (5.33) imply that

$$\sup(\pi^u) \geq \pi^u(x_0) \geq \limsup_{k \to \infty} \pi^u\left(x_0^{(k)}\right) \geq \sup(\pi^u). \tag{5.37}$$

By upper semicontinuity of u, Proposition 5.7, (5.32)–(5.34), (5.36), and (5.37),

$$\sum_{t=0}^{T_0-1} (u(x_t, x_{t+1}) - u(\bar{x}, \bar{x})) - \pi^u(x_0) + \pi^u(x_{T_0})$$

$$\geq \limsup_{k \to \infty} \left(\sum_{t=0}^{T_0-1} (u(x_t^{(k)}, x_{t+1}^{(k)}) - u(\bar{x}, \bar{x})) - \pi^u(x_0^{(k)}) + \pi^u(x_{T_0}^{(k)}) \right)$$

$$\geq \limsup_{k \to \infty}(-\delta_k) = 0.$$

Thus

$$\sum_{t=0}^{T_0-1} (u(x_t, x_{t+1}) - u(\bar{x}, \bar{x})) - \pi^u(x_0) + \pi^u(x_{T_0}) \geq 0.$$

Together with Proposition 5.1 this implies that for all integers $t = 0, \ldots, T_0 - 1$,

$$u(x_t, x_{t+1}) - u(\bar{x}, \bar{x}) = \pi^u(x_t) - \pi^u(x_{t+1}). \tag{5.38}$$

Theorem 3.4 implies that there is a (u, Ω)-overtaking optimal program $\{\tilde{x}_t\}_{t=0}^{\infty}$ satisfying

$$\tilde{x}_0 = x_{T_0}. \tag{5.39}$$

For all integers $t > T_0$ set

$$x_t = \tilde{x}_{t-T_0}. \tag{5.40}$$

It is clear that $\{x_t\}_{t=0}^{\infty}$ is an (Ω)-program. Since $\{\tilde{x}_t\}_{t=0}^{\infty}$ is a (u, Ω)-overtaking optimal program, Corollary 5.3 implies that (5.38) holds for all integers $t \geq 0$. It follows from (5.38) and Proposition 5.8 that $\{x_t\}_{t=0}^{\infty}$ is a (u, Ω)-overtaking optimal program. In view of (5.37),

$$\pi^u(x_0) = \sup(\pi^u).$$

By (5.36), for all sufficiently large natural numbers k, $\rho(x_t, x_t^{(k)}) \leq \epsilon/4, t = 0, \ldots, T_0$. This contradicts (5.35). The contradiction we have reached proves Lemma 5.14.

5.4   Proof of Theorem 5.13

We may assume that $\epsilon < 1$. By Lemma 5.14, there exists $\delta_1 \in (0, \epsilon/4)$ such that the following property holds:

(i) for each (Ω)-program $\{y_t\}_{t=0}^{T_0}$ which satisfies

$$\pi^u(y_0) \geq \sup(\pi^u) - \delta_1,$$

$$\sum_{t=0}^{T_0-1} (u(y_t, y_{t+1}) - u(\bar{x}, \bar{x})) - \pi^u(y_0) + \pi^u(y_{T_0}) \geq -\delta_1$$

there exists a (u, Ω)-overtaking program $\{z_t\}_{t=0}^{\infty}$ such that

$$\pi^u(z_0) = \sup(\pi^u),$$

$$\rho(z_t, y_t) \leq \epsilon, \ t = 0, \ldots, T_0.$$

By Propositions 5.4 and 5.5 and (B2), there exists $\delta_2 \in (0, \delta_1)$ such that the following properties hold:

(ii) for each $z \in K$ satisfying $\rho(z, \bar{x}) \leq 2\delta_2$, we have

$$|\pi^u(z)| = |\pi^u(z) - \pi^u(\bar{x})| \leq \delta_1/8;$$

(iii) for each $(x, y) \in \Omega$ satisfying

$$\rho(x, \bar{x}) \le 2\delta_2, \rho(y, \bar{x}) \le 2\delta_2$$

we have

$$|u(x, y) - u(\bar{x}, \bar{x})| \le (\delta_1/8)(L_* + 1)^{-1}.$$

By (B3), there exists $\delta_3 \in (0, \delta_2)$ such that the following property holds:
(iv) for each $x, y \in K$ satisfying

$$\rho(x, \bar{x}) \le \delta_3, \ \rho(y, \bar{x}) \le \delta_3$$

there exists an (Ω)-program $\{\xi_t\}_{t=0}^{L_*}$ such that

$$\xi_0 = x, \ \xi_{L_*} = y$$

and

$$\rho(\xi_t, \bar{x}) \le \delta_2, \ t = 0, \dots, L_*.$$

By Theorem 3.12, there exist an integer $L_0 \ge 1$ and a number $\delta_4 > 0$ such that the following property holds:
(v) for each integer $T > 2L_0$, each finite sequence of bounded functions $u_t : \Omega \to R^1, t = 0, \dots T - 1$ satisfying

$$\|u_t - u\| \le \delta_4, \ t = 0 \dots, T - 1$$

and each (Ω)-program $\{z_t\}_{t=0}^T$ which satisfies

$$\sum_{t=0}^{T-1} u_t(z_t, z_{t+1}) \ge \widehat{U}\left(\{u_t\}_{t=0}^{T-1}, \Omega, z_T\right) - \delta_4$$

we have

$$\rho(z_t, \bar{x}) \le \delta_3, \ t = L_0, \dots, T - L_0.$$

By Theorem 3.4, there exists a (u, Ω)-overtaking optimal program $\{z_t\}_{t=0}^\infty$ such that

$$\pi^u(z_0) = \sup(\pi^u). \tag{5.41}$$

(B1) implies that there exists a natural number τ_1 such that

$$\rho(z_t, \bar{x}) \le \delta_3 \text{ for all integers } t \ge \tau_1. \tag{5.42}$$

Choose a positive number δ and an integer T_0 such that

$$\delta < (16(L_0 + L_* + \tau_1 + \tau_0 + 6))^{-1} \min\{\delta_1, \delta_2, \delta_3, \delta_4\}, \tag{5.43}$$

$$T_0 > 2L_* + 2L_0 + 2\tau_0 + 2\tau_1 + 4. \tag{5.44}$$

Assume that an integer $T \geq T_0$, a finite sequence of bounded functions $u_t : \Omega \to R^1, t = 0, \ldots T - 1$ satisfies

$$\|u_t - u\| \leq \delta, \ t = 0 \ldots, T - 1 \tag{5.45}$$

and $\{x_t\}_{t=0}^T$ is an (Ω)-program which satisfies

$$\sum_{t=0}^{T-1} u_t(x_t, x_{t+1}) \geq U\left(\{u_t\}_{t=0}^{T-1}, \Omega, x_T\right) - \delta. \tag{5.46}$$

By (5.43)–(5.46) and the property (v),

$$\rho(x_t, \bar{x}) \leq \delta_3, \ t = L_0, \ldots, T - L_0. \tag{5.47}$$

Property (iv), (5.42), (5.44), and (5.47) imply that there exist an (Ω)-program $\{\tilde{x}_t\}_{t=0}^T$ such that

$$\tilde{x}_t = z_t, \ t = 0, \ldots, L_0 + \tau_0 + \tau_1 + 3, \tag{5.48}$$

$$\rho(\tilde{x}_t, \bar{x}) \leq \delta_2, \ t = L_0 + \tau_0 + \tau_1 + 4, \ldots, L_0 + L_* + \tau_0 + \tau_1 + 3, \tag{5.49}$$

$$\tilde{x}_t = x_t, \ t = L_0 + L_* + \tau_0 + \tau_1 + 3, \ldots, T. \tag{5.50}$$

It follows from (5.45), (5.46), and (5.50) that

$$\delta \geq \sum_{t=0}^{T-1} (u_t(\tilde{x}_t, \tilde{x}_{t+1}) - u_t(x_t, x_{t+1}))$$

$$= \sum_{t=0}^{L_0+L_*+\tau_0+\tau_1+2} (u_t(\tilde{x}_t, \tilde{x}_{t+1}) - u_t(x_t, x_{t+1}))$$

$$\geq \sum_{t=0}^{L_0+L_*+\tau_0+\tau_1+2} (u(\tilde{x}_t, \tilde{x}_{t+1}) - u(x_t, x_{t+1}))$$

$$- 2\delta(L_0 + L_* + \tau_0 + \tau_1 + 3). \tag{5.51}$$

Properties (ii) and (iii), (5.42), (5.48), and (5.49) imply that

$$|u(\tilde{x}_t, \tilde{x}_{t+1}) - u(\bar{x}, \bar{x})| \le (\delta_1/8)(L_* + 1)^{-1}, \tag{5.52}$$

$$t = L_0 + \tau_0 + \tau_1 + 3, \dots, L_0 + L_* + \tau_0 + \tau_1 + 2.$$

Corollary 5.3, property (ii), (5.42), (5.43), (5.44), (5.47), (5.48), (5.51), and (5.52) imply that

$$\delta \ge \sum_{t=0}^{L_0+L_*+\tau_0+\tau_1+2} u(z_t, z_{t+1}) + L_* u(\bar{x}, \bar{x}) - L_*(\delta_1/8)(L_* + 1)^{-1}$$

$$- \sum_{t=0}^{L_0+L_*+\tau_0+\tau_1+2} u(x_t, x_{t+1}) - 2\delta(L_0 + L_* + \tau_0 + \tau_1 + 3)$$

$$= \pi^u(z_0) - \pi^u(z_{L_0+\tau_0+\tau_1+3})$$

$$+ (L_0 + L_* + \tau_0 + \tau_1 + 3)u(\bar{x}, \bar{x}) - \delta_1/4$$

$$- \left(\sum_{t=0}^{L_0+L_*+\tau_1+2} (u(x_t, x_{t+1}) - u(\bar{x}, \bar{x})) - \pi^u(x_0) + \pi^u(x_{L_0+L_*+\tau_1+3}) \right)$$

$$- (L_0 + L_* + \tau_1 + 3)u(\bar{x}, \bar{x}) - \pi^u(x_0) + \pi^u(x_{L_0+L_*+\tau_1+3})$$

$$\ge \sup(\pi^u) - \delta_1/8 - \delta_1/4 - \pi^u(x_0) - \delta_1/8$$

$$- \left(\sum_{t=0}^{L_0+L_*+\tau_1+2} (u(x_t, x_{t+1}) - u(\bar{x}, \bar{x})) - \pi^u(x_0) + \pi^u(x_{L_0+L_*+\tau_1+3}) \right). \tag{5.53}$$

It follows from Proposition 5.1, (5.43), and (5.53) that

$$\pi^u(x_0) + \sum_{t=0}^{\tau_0-1} (u(x_t, x_{t+1}) - u(\bar{x}, \bar{x})) - \pi^u(x_0) + \pi^u(x_{\tau_0})$$

$$\ge \sup(\pi^u) - \delta_1,$$

$$\pi^u(x_0) \ge \sup(\pi^u) - \delta_1, \tag{5.54}$$

$$\sum_{t=0}^{\tau_0-1} (u(x_t, x_{t+1}) - u(\bar{x}, \bar{x})) - \pi^u(x_0) + \pi^u(x_{\tau_0}) \ge -\delta_1. \tag{5.55}$$

By (5.54), (5.55), and the property (i), there exists a (u, Ω)-overtaking optimal program $\{y_t\}_{t=0}^{\infty}$ such that

$$\pi^u(y_0) = \sup(\pi^u),$$

$$\rho(y_t, x_t) \le \epsilon, \ t = 0, \ldots, \tau_0.$$

Theorem 5.13 is proved.

5.5 The First Class of Bolza Problems

For each nonempty set Y and each function $h : Y \to R^1$ put

$$\sup(h) = \sup\{h(y) : \ y \in Y\}.$$

Denote by $\mathcal{M}(K)$ the set of all bounded functions $h : K \to R^1$. For each $h \in \mathcal{M}(K)$ set

$$\|h\| = \sup\{|h(x)| : \ x \in K\}.$$

Clearly, $(\mathcal{M}(K), \| \cdot \|)$ is a Banach space. For each $h_1, h_2 \in \mathcal{M}(K)$ set

$$d_K(h_1, h_2) = \|h_1 - h_2\|.$$

For each $x \in K$, each pair of integers T_1, T_2 satisfying $0 \le T_1 < T_2$, each finite sequence of bounded functions $u_t : \Omega \to R^1, t = T_1, \ldots, T_2 - 1$, and each $h \in \mathcal{M}(K)$ we consider the problem

$$\sum_{t=T_1}^{T_2-1} u_t(x_t, x_{t+1}) + h(x_{T_2}) \to \max, \ \{(x_t, x_{t+1})\}_{t=0}^{T-1} \subset \Omega, \ x_{T_1} = x \qquad (P_{B,1})$$

and set

$$U\left(h, \{u_t\}_{t=T_1}^{T_2-1}, \Omega, x\right) = \sup \left\{ \sum_{t=T_1}^{T_2-1} u_t(x_t, x_{t+1}) + h(x_{T_2}) : \right.$$

$$\left. \{x_t\}_{t=T_1}^{T_2} \text{ is an } (\Omega) - \text{program and } x_{T_1} = x \right\}.$$

For each $x \in X$, each pair of integers T_1, T_2 satisfying $0 \le T_1 < T_2$, each bounded function $u : \Omega \to R^1$, and each $h \in \mathcal{M}(K)$ set

$$U(h, u, T_1, T_2, \Omega, x) = U(h, \{u_t\}_{t=T_1}^{T_2-1}, \Omega, x) \text{ where } u_t = u, \ t = T_1, \ldots, T_2 - 1.$$

In Sect. 5.7 we prove the following result which describes the structure of approximate solutions of the problems of the type $(P_{B,1})$ in the regions close to the right endpoints.

Theorem 5.15 *Let $g \in \mathcal{M}(K)$ be upper semicontinuous function, $\tau_0 \geq 1$ be an integer, and $\epsilon > 0$. Then there exist $\delta > 0$ and an integer $T_0 \geq \tau_0$ such that for each integer $T \geq T_0$, each $h \in \mathcal{M}(K)$ satisfying*

$$\|h - g\| \leq \delta,$$

each finite sequence of bounded functions $u_t : \Omega \to R^1, t = 0, \ldots, T-1$ satisfying

$$\|u_t - u\| \leq \delta, \ t = 0 \ldots, T - 1$$

and each (Ω)-program $\{x_t\}_{t=0}^T$ which satisfies

$$\sum_{t=\tau}^{\tau+T_0-1} u_t(x_t, x_{t+1}) \geq U\left(\{u_t\}_{t=\tau}^{\tau+T_0-1}, \Omega, x_\tau, x_{\tau+T_0}\right) - \delta$$

for each $\tau \in \{0, \ldots, T - T_0\}$,

$$h(x_T) + \sum_{t=T-T_0}^{T-1} u_t(x_t, x_{t+1}) \geq U\left(h, \{u_t\}_{t=T-T_0}^{T-1}, \Omega, x_{T-T_0}\right) - \delta$$

there exists a $(\bar{u}, \bar{\Omega})$-overtaking optimal program $\{x_t^*\}_{t=0}^\infty$ such that

$$(\pi^{\bar{u}} + g)(x_0^*) = \sup(\pi^{\bar{u}} + g),$$

$$\rho(x_{T-t}, x_t^*) \leq \epsilon, \ t = 0, \ldots, \tau_0.$$

Let $g \in \mathcal{M}(K)$ be upper semicontinuous. Then $\pi^{\bar{u}} + g : K \to R^1$ is an upper semicontinuous, bounded function. Therefore there exists $x \in K$ such that

$$(\pi^{\bar{u}} + g)(x) = \sup(\pi^{\bar{u}} + g).$$

5.6 An Auxiliary Result for Theorem 5.15

Lemma 5.16 *Let $g \in \mathcal{M}(K)$ be an upper semicontinuous function, $T_0 \geq 1$ be an integer, and $\epsilon \in (0, 1)$. Then there exists $\delta \in (0, \epsilon)$ such that for each (Ω)-program $\{x_t\}_{t=0}^{T_0}$ which satisfies*

$$(\pi^u + g)(x_0) \geq \sup(\pi^u + g) - \delta,$$

$$\sum_{t=0}^{T_0-1} (u(x_t, x_{t+1}) - u(\bar{x}, \bar{x})) - \pi^u(x_0) + \pi^u(x_{T_0}) \geq -\delta$$

there exists a (u, Ω)-overtaking optimal program $\{z_t\}_{t=0}^{\infty}$ such that

$$(\pi^u + g)(z_0) = \sup(\pi^u + g), \tag{5.56}$$

$$\rho(z_t, x_t) \leq \epsilon, \quad t = 0, \ldots, T_0.$$

Proof Assume that the lemma does not hold. Then there exist a sequence of real numbers $\{\delta_k\}_{k=1}^{\infty} \subset (0, 1]$ and a sequence of (Ω)-programs $\{x_t^{(k)}\}_{t=0}^{T_0}, k = 1, 2, \ldots$ such that

$$\lim_{k \to \infty} \delta_k = 0 \tag{5.57}$$

and that for each natural number k and each (u, Ω)-overtaking optimal program $\{z_t\}_{t=0}^{\infty}$ satisfying (5.56),

$$(\pi^u + g)\left(x_0^{(k)}\right) \geq \sup(\pi^u + g) - \delta_k, \tag{5.58}$$

$$\sum_{t=0}^{T_0-1} \left(u\left(x_t^{(k)}, x_{t+1}^{(k)}\right) - u(\bar{x}, \bar{x})\right) - \pi^u\left(x_0^{(k)}\right) + \pi^u\left(x_{T_0}^{(k)}\right) \geq -\delta_k, \tag{5.59}$$

we have

$$\max\left\{\rho(z_t, x_t^{(k)}) : t = 0, \ldots, T_0\right\} > \epsilon. \tag{5.60}$$

Extracting a subsequence and re-indexing, if necessary, we may assume without loss of generality that for each integer $t \in [0, T_0]$ there exists

$$x_t = \lim_{k \to \infty} x_t^{(k)}. \tag{5.61}$$

By (5.61) and upper semicontinuity of g and π^u,

$$\pi^u(x_0) \geq \limsup_{k \to \infty} \pi^u(x_0^{(k)}),$$

$$g(x_0) \geq \limsup_{k \to \infty} g(x_0^{(k)}). \tag{5.62}$$

It follows from (5.57), (5.58), and (5.62) that

$$\sup(\pi^u + g) \geq (\pi^u + g)(x_0)$$

$$\geq \limsup_{k \to \infty}(\pi^u + g)(x_0^{(k)}) \geq \sup(\pi^u + g),$$

$$\sup(\pi^u + g) = (\pi^u + g)(x_0) = \lim_{k \to \infty}(\pi^u + g)(x_0^{(k)}). \tag{5.63}$$

Relations (5.62) and (5.63) imply that

$$\pi^u(x_0) = \lim_{k \to \infty} \pi^u(x_0^{(k)}),$$

$$g(x_0) = \lim_{k \to \infty} g(x_0^{(k)}). \tag{5.64}$$

It follows from upper semicontinuity of u and π^u, (5.57), (5.59), (5.61), and (5.64) that

$$\sum_{t=0}^{T_0-1}(u(x_t, x_{t+1}) - u(\bar{x}, \bar{x})) - \pi^u(x_0) + \pi^u(x_{T_0})$$

$$\geq \limsup_{k \to \infty}\left(\sum_{t=0}^{T_0-1}(u(x_t^{(k)}, x_{t+1}^{(k)}) - u(\bar{x}, \bar{x})) - \pi^u(x_0^{(k)}) + \pi^u(x_{T_0}^{(k)})\right)$$

$$\geq \limsup_{k \to \infty}(-\delta_k) = 0.$$

Combined with Proposition 5.1 this implies that for all integers $t = 0, \ldots, T_0 - 1$,

$$u(x_t, x_{t+1}) - u(\bar{x}, \bar{x}) = \pi^u(x_t) - \pi^u(x_{t+1}). \tag{5.65}$$

It follows from Theorem 3.4 that there exists a (u, Ω)-overtaking optimal program $\{\tilde{x}_t\}_{t=0}^{\infty}$ satisfying

$$\tilde{x}_0 = x_{T_0}. \tag{5.66}$$

For all integers $t > T_0$ set

$$x_t = \tilde{x}_{t-T_0}. \tag{5.67}$$

Evidently, $\{x_t\}_{t=0}^{\infty}$ is an (Ω)-program. It follows from Corollary 5.3 that (5.65) holds for all integers $t \geq 0$. Proposition 5.8 and (5.65) imply that $\{x_t\}_{t=0}^{\infty}$ is a (u, Ω)-overtaking optimal program satisfying (5.63). By (5.61), for all sufficiently large natural numbers k, $\rho(x_t, x_t^{(k)}) \leq \epsilon/4$, $t = 0, \ldots, T_0$. This contradicts (5.60). The contradiction we have reached proves Lemma 5.16.

5.7 Proof of Theorem 5.15

By Lemma 5.16 applied to the pair $(\bar{u}, \bar{\Omega})$ there exists a real number

$$\delta_1 \in (0, \epsilon/2)$$

such that the following property holds:

(i) for each $(\bar{\Omega})$-program $\{y_t\}_{t=0}^{\tau_0}$ which satisfies

$$(\pi^{\bar{u}} + g)(y_0) \geq \sup(\pi^{\bar{u}} + g) - \delta_1,$$

$$\sum_{t=0}^{\tau_0-1} (\bar{u}(y_t, y_{t+1}) - \bar{u}(\bar{x}, \bar{x})) - \pi^{\bar{u}}(y_0) + \pi^{\bar{u}}(y_{\tau_0}) \geq -\delta_1$$

there exists an $(\bar{u}, \bar{\Omega})$-overtaking optimal program $\{z_t\}_{t=0}^{\infty}$ such that

$$(\pi^{\bar{u}} + g)(z_0) = \sup(\pi^{\bar{u}} + g),$$

$$\rho(z_t, y_t) \leq \epsilon, \ t = 0, \ldots, \tau_0.$$

Propositions 5.4 and 5.5 applied to the pair $(\bar{u}, \bar{\Omega})$ and assumption (B2) imply that there exists a real number $\delta_2 \in (0, \delta_1)$ such that the following properties hold:

(ii) for each point $z \in K$ satisfying $\rho(z, \bar{x}) \leq 2\delta_2$,

$$|\pi^{\bar{u}}(z)| = |\pi^{\bar{u}}(z) - \pi^{\bar{u}}(\bar{x})| \leq \delta_1/16;$$

(iii) for each $(x, y) \in \Omega$ satisfying

$$\rho(x, \bar{x}) \leq 2\delta_2, \ \rho(y, \bar{x}) \leq 2\delta_2,$$

$$|u(x, y) - u(\bar{x}, \bar{x})| \leq (\delta_1/8)(L_* + 1)^{-1}.$$

By (B3), there exists $\delta_3 \in (0, \delta_2)$ such that the following property holds:

(iv) for each $x, y \in K$ satisfying

$$\rho(x, \bar{x}) \leq \delta_3, \ \rho(y, \bar{x}) \leq \delta_3$$

there exists an (Ω)-program $\{\xi_t\}_{t=0}^{L_*}$ such that

$$\xi_0 = x, \ \xi_{L_*} = y$$

and

$$\rho(\xi_t, \bar{x}) \leq \delta_2, \ t = 0, \ldots, L_*.$$

By Theorem 3.12, there exist an integer $L_1 \geq 1$ and a number $\delta_4 \in (0, \delta_3)$ such that the following property holds:

(v) for each integer $T > 2L_1$, each finite sequence of bounded functions $u_t : \Omega \to R^1, t = 0, \ldots T - 1$ satisfying

$$\|u_t - u\| \leq \delta_4, \ t = 0 \ldots, T - 1$$

and each (Ω)-program $\{z_t\}_{t=0}^{T}$ which satisfies

$$\sum_{t=\tau}^{\tau+L_1-1} u_t(z_t, z_{t+1}) \geq U\left(\{u_t\}_{t=\tau}^{\tau+L_1-1}, \Omega, z_\tau, z_{\tau+L_1}\right) - \delta_4$$

for each integer $\tau \in [0, T - L_1]$, we have

$$\rho(z_t, \bar{x}) \leq \delta_3, \ t = L_1, \ldots, T - L_1.$$

By Theorem 3.4, there exists a $(\bar{u}, \bar{\Omega})$-overtaking optimal program $\{z_t\}_{t=0}^{\infty}$ such that

$$(\pi^{\bar{u}} + g)(z_0) = \sup(\pi^{\bar{u}} + g). \tag{5.68}$$

(B1) implies that there exists a natural number τ_1 such that

$$\rho(z_t, \bar{x}) \leq \delta_3 \text{ for all integers } t \geq \tau_1. \tag{5.69}$$

Choose a positive number δ and an integer T_0 such that

$$\delta < (16(L_1 + L_* + \tau_1 + \tau_0 + 8))^{-1} \min\{\delta_1, \delta_2, \delta_3, \delta_4\}, \tag{5.70}$$

$$T_0 > 2L_* + 2L_1 + 2\tau_0 + 2\tau_1 + 8. \tag{5.71}$$

Assume that an integer $T \geq T_0, h \in \mathcal{M}(K)$ satisfies

$$\|h - g\| \leq \delta, \tag{5.72}$$

a finite sequence of bounded functions $u_t : \Omega \to R^1, t = 0, \ldots T - 1$ satisfies

$$\|u_t - u\| \leq \delta, \ t = 0 \ldots, T - 1 \tag{5.73}$$

and $\{x_t\}_{t=0}^{T}$ is an (Ω)-program which satisfies

$$\sum_{t=\tau}^{\tau+T_0-1} u_t(x_t, x_{t+1}) \geq U\left(\{u_t\}_{t=\tau}^{\tau+T_0-1}, \Omega, x_\tau, x_{\tau+T_0}\right) - \delta \tag{5.74}$$

for each integer $\tau \in [0, T - T_0]$,

$$h(x_T) + \sum_{t=T-T_0}^{T-1} u_t(x_t, x_{t+1}) \geq U\left(h, \{u_t\}_{t=T-T_0}^{T-1}, \Omega, x_{T-T_0}\right) - \delta. \tag{5.75}$$

By (5.70), (5.71), (5.73), (5.74), and the property (v),

$$\rho(x_t, \bar{x}) \leq \delta_3, \ t = L_1, \ldots, T - L_1. \tag{5.76}$$

Property (iv), (5.69), (5.71), and (5.76) imply that there exists an (Ω)-program $\{\tilde{x}_t\}_{t=0}^{T}$ such that

$$\tilde{x}_t = x_t, \ t = 0, \ldots, T - L_1 - L_* - \tau_0 - \tau_1 - 4, \tag{5.77}$$

$$\tilde{x}_t = z_{T-t}, \ t = T - L_1 - \tau_0 - \tau_1 - 4, \ldots, T, \tag{5.78}$$

$$\rho(\tilde{x}_t, \bar{x}) \leq \delta_2, \ t = T - L_1 - L_* - \tau_0 - \tau_1 - 4, \ldots, T - L_1 - \tau_0 - \tau_1 - 4. \tag{5.79}$$

By property (iii), (5.71)–(5.73), (5.75), and (5.77),

$$\delta \geq U\left(h, \{u_t\}_{t=T-T_0}^{T-1}, \Omega, x_{T-T_0}\right) - h(x_T) - \sum_{t=T-T_0}^{T-1} u_t(x_t, x_{t+1})$$

$$\geq h(\tilde{x}_T) + \sum_{t=T-T_0}^{T-1} u_t(\tilde{x}_t, \tilde{x}_{t+1}) - h(x_T) - \sum_{t=T-T_0}^{T-1} u_t(x_t, x_{t+1})$$

$$= h(\tilde{x}_T) - h(x_T) + \sum_{t=T-L_1-L_*-\tau_0-\tau_1-4}^{T-1} (u_t(\tilde{x}_t, \tilde{x}_{t+1}) - u_t(x_t, x_{t+1}))$$

$$\geq g(\tilde{x}_T) - g(x_T) + \sum_{t=T-L_1-L_*-\tau_0-\tau_1-4}^{T-1} u(\tilde{x}_t, \tilde{x}_{t+1})$$

$$\quad - \sum_{t=T-L_1-L_*-\tau_0-\tau_1-4}^{T-1} u(x_t, x_{t+1}) - 2\delta(L_1 + L_* + \tau_0 + \tau_1 + 5)$$

$$\geq g(\tilde{x}_T) - g(x_T)$$

$$+ \sum_{t=T-L_1-\tau_0-\tau_1-4}^{T-1} u(\tilde{x}_t, \tilde{x}_{t+1}) + L_* u(\bar{x}, \bar{x}) - L_*(\delta_1/8)(L_* + 1)^{-1}$$

$$- \sum_{t=T-L_1-L_*-\tau_0-\tau_1-4}^{T-1} u(x_t, x_{t+1}) - 2\delta(L_1 + L_* + \tau_0 + \tau_1 + 5).$$

Combined with (5.22), (5.70), and (5.78) this implies that

$$
g(x_T) + \sum_{t=T-L_1-L_*-\tau_0-\tau_1-4}^{T-1} u(x_t, x_{t+1})
$$

$$
\geq g(\tilde{x}_T) + \sum_{t=T-L_1-\tau_0-\tau_1-4}^{T-1} u(\tilde{x}_t, \tilde{x}_{t+1}) + L_* u(\bar{x}, \bar{x}) - \delta_1/8 - \delta_1/8
$$

$$
= g(z_0) + \sum_{t=0}^{L_1+\tau_0+\tau_1+2} \bar{u}(z_t, z_{t+1}) + u(\tilde{x}_{T-L_1-\tau_0-\tau_1-4}, \tilde{x}_{T-L_1-\tau_0-\tau_1-3})
$$

$$
+ L_* u(\bar{x}, \bar{x}) - \delta_1/4. \tag{5.80}
$$

Property (iii), (5.69), and (5.78) imply that

$$
|u(\tilde{x}_{T-L_1-\tau_0-\tau_1-4}, \tilde{x}_{T-L_1-\tau_0-\tau_1-3}) - u(\bar{x}, \bar{x})| \leq \delta_1/8. \tag{5.81}
$$

In view of (5.80) and (5.81),

$$
g(x_T) + \sum_{t=T-L_1-L_*-\tau_0-\tau_1-4}^{T-1} u(x_t, x_{t+1})
$$

$$
\geq g(z_0) + \sum_{t=0}^{L_1+\tau_0+\tau_1+2} \bar{u}(z_t, z_{t+1}) + (L_* + 1)u(\bar{x}, \bar{x}) - 3\delta_1/8. \tag{5.82}
$$

Set

$$
y_t = x_{T-t}, \quad t = 0, \ldots, T. \tag{5.83}
$$

By (5.22), (5.82), and (5.83),

$$
g(y_0) + \sum_{t=0}^{L_1+L_*+\tau_0+\tau_1+3} \bar{u}(y_t, y_{t+1})
$$

$$
= g(x_T) + \sum_{T-L_1-L_*-\tau_0-\tau_1-4}^{T-1} u(x_t, x_{t+1})
$$

$$
\geq g(z_0) + \sum_{t=0}^{L_1+\tau_0+\tau_1+2} \bar{u}(z_t, z_{t+1}) + (L_* + 1)u(\bar{x}, \bar{x}) - 3\delta_1/8. \tag{5.84}
$$

By (5.68), (5.84), Proposition 5.1, and Corollary 5.3,

$$(\pi^{\bar{u}} + g)(y_0) - \sup(\pi^{\bar{u}} + g)$$

$$+ \sum_{t=0}^{\tau_0-1} (\bar{u}(y_t, y_{t+1}) - \bar{u}(\bar{x}, \bar{x})) - \pi^{\bar{u}}(y_0) + \pi^{\bar{u}}(y_{\tau_0})$$

$$\geq (\pi^{\bar{u}} + g)(y_0) - (\pi^{\bar{u}} + g)(z_0)$$

$$+ \sum_{t=0}^{L_1+L_*+\tau_0+\tau_1+3} (\bar{u}(y_t, y_{t+1}) - u(\bar{x}, \bar{x})) - \pi^{\bar{u}}(y_0) + \pi^{\bar{u}}(y_{L_1+L_*+\tau_0+\tau_1+4})$$

$$\geq \pi^{\bar{u}}(y_0) - \pi^{\bar{u}}(z_0) + \sum_{t=0}^{L_1+\tau_0+\tau_1+2} (\bar{u}(z_t, z_{t+1}) - \bar{u}(\bar{x}, \bar{x}))$$

$$- 3\delta_1/8 - \pi^{\bar{u}}(y_0) + \pi^{\bar{u}}(y_{L_1+L_*+\tau_0+\tau_1+4})$$

$$= -\pi^{\bar{u}}(z_{L_1+\tau_0+\tau_1+3}) - \pi^{\bar{u}}(y_{L_1+L_*+\tau_0+\tau_1+4}) - 3\delta_1/8. \tag{5.85}$$

Property (ii), (5.69), (5.71), (5.76), and (5.83) imply that

$$|\pi^{\bar{u}}(z_{L_1+\tau_0+\tau_1+3})| \leq \delta_1/8, \tag{5.86}$$

$$|\pi^{\bar{u}}(y_{L_1+L_*+\tau_0+\tau_1+4})| \leq \delta_1/8. \tag{5.87}$$

It follows from (5.85)–(5.87) that

$$(\pi^{\bar{u}} + g)(y_0) - \sup(\pi^{\bar{u}} + g)$$

$$+ \sum_{t=0}^{\tau_0-1} (\bar{u}(y_t, y_{t+1}) - \bar{u}(\bar{x}, \bar{x}))$$

$$- \pi^{\bar{u}}(y_0) + \pi^{\bar{u}}(y_{\tau_0}) \geq -\delta_1.$$

Together with Proposition 5.1 this implies that

$$(\pi^{\bar{u}} + g)(y_0) - \sup(\pi^{\bar{u}} + g) \geq -\delta_1,$$

$$\sum_{t=0}^{\tau_0-1} (\bar{u}(y_t, y_{t+1}) - \bar{u}(\bar{x}, \bar{x}))$$

$$- \pi^{\bar{u}}(y_0) + \pi^{\bar{u}}(y_{\tau_0}) \geq -\delta_1.$$

It follows from the relations above, (5.83) and the property (i) that there exists an $(\bar{u}, \bar{\Omega})$-overtaking optimal program $\{\xi_t\}_{t=0}^{\infty}$ such that

$$(\pi^{\bar{u}} + g)(\xi_0) = \sup(\pi^{\bar{u}} + g),$$

$\rho(\xi_t, x_{T-t}) = \rho(\xi_t, y_t) \leq \epsilon$, $t = 0, \ldots, \tau_0$. Theorem 5.15 is proved.

5.8   The Second Class of Bolza Problems

Denote by $\mathcal{M}(K \times K)$ the collection of all bounded functions $h : K \times K \to R^1$. For every function $h \in \mathcal{M}(K \times K)$ put

$$\|h\| = \sup\{|h(x, y)| : x, y \in K\}.$$

It is not difficult to see that $(\mathcal{M}(K \times K), \| \cdot \|)$ is a Banach space.

For every pair of nonnegative integers $T_1 < T_2$, every finite sequence of bounded functions $u_t : \Omega \to R^1$, $t = T_1, \ldots, T_2 - 1$, and every function $h \in \mathcal{M}(K \times K)$ we consider the problem

$$\sum_{t=T_1}^{T_2-1} u_t(x_t, x_{t+1}) + h(x_{T_1}, x_{T_2}) \to \max, \ \{(x_t, x_{t+1})\}_{t=T_1}^{T_2-1} \subset \Omega$$

and define

$$U\left(h, \{u_t\}_{t=T_1}^{T_2-1}, \Omega\right) = \sup\left\{ \sum_{t=T_1}^{T_2-1} u_t(x_t, x_{t+1}) + h(x_{T_1}, x_{T_2}) : \right.$$

$$\left. \{x_t\}_{t=T_1}^{T_2} \text{ is an } (\Omega) - \text{program} \right\}.$$

For every pair of nonnegative integers $T_1 < T_2$, every bounded function $v : \Omega \to R^1$, and every function $h \in \mathcal{M}(K \times K)$ define

$$U(h, v, \Omega, T_1, T_2) = U(h, \{v_t\}_{t=T_1}^{T_2-1}, \Omega) \text{ where } v_t = v, \ t = T_1, \ldots, T_2 - 1.$$

Let $g \in \mathcal{M}(K \times K)$ be an upper semicontinuous function. Then the function

$$\psi_g(\xi, \eta) = \pi^u(\xi) + \pi^{\bar{u}}(\eta) + g(\xi, \eta), \ (\xi, \eta) \in \Omega \tag{5.88}$$

is upper semicontinuous bounded function which has a point of minimum.

Our next result describes the structure of approximate solutions in the regions close to the endpoints.

Theorem 5.17 *Suppose that $g \in \mathcal{M}(K \times K)$ is an upper semicontinuous function, τ_0 is a natural number, and $\epsilon \in (0, 1)$. Then there exist a positive number δ and a natural number $T_0 \geq \tau_0$ such that for every natural number $T \geq T_0$, every function $h \in \mathcal{M}(K \times K)$ satisfying $\|h - g\| \leq \delta$, every finite sequence of bounded functions $u_t : \Omega \to R^1, t = 0, \ldots, T - 1$ which satisfies*

$$\|u_t - u\| \leq \delta, \ t = 0 \ldots, T - 1,$$

and every (Ω)-program $\{x_t\}_{t=0}^{T}$ which satisfies

$$h(x_0, x_T) + \sum_{t=0}^{T-1} u_t(x_t, x_{t+1}) \geq U\left(h, \{u_t\}_{t=0}^{T-1}, \Omega\right) - \delta,$$

there exist a (u, Ω)-overtaking optimal program $\{x_t^*\}_{t=0}^{\infty}$ and a $(\bar{u}, \bar{\Omega})$-overtaking optimal program $\{\bar{x}_t^*\}_{t=0}^{\infty}$ such that

$$\pi^u(x_0^*) + \pi^{\bar{u}}(\bar{x}_0^*) + g(x_0^*, \bar{x}_0^*)$$

$$\geq \pi^u(\xi) + \pi^{\bar{u}}(\eta) + g(\xi, \eta) \text{ for all } \xi, \eta \in K$$

and that for all integers $t = 0, \ldots, \tau_0$, $\rho(x_t, x_t^*) \leq \epsilon$ and $\rho(x_{T-t}, \bar{x}_t^*) \leq \epsilon$.

5.9 Auxiliary Results for Theorem 5.17

Lemma 5.18 *Suppose that $g \in \mathcal{M}(K \times K)$ is an upper semicontinuous function, T_0 is a natural number, and $\epsilon \in (0, 1)$. Then there exists $\delta \in (0, \epsilon)$ such that for every (Ω)-program $\{x_t\}_{t=0}^{T_0}$ and every $(\bar{\Omega})$-program $\{y_t\}_{t=0}^{T_0}$ which satisfy*

$$\psi_g(x_0, y_0) + \delta \geq \sup(\psi_g),$$

$$\sum_{t=0}^{T_0-1} (u(x_t, x_{t+1}) - u(\bar{x}, \bar{x})) - \pi^u(x_0) + \pi^u(x_{T_0}) \geq -\delta,$$

$$\sum_{t=0}^{T_0-1} (\bar{u}(y_t, y_{t+1}) - u(\bar{x}, \bar{x})) - \pi^{\bar{u}}(y_0) + \pi^{\bar{u}}(y_{T_0}) \geq -\delta,$$

there exist a (u, Ω)-overtaking optimal program $\{x_t^*\}_{t=0}^{\infty}$ and an $(\bar{u}, \bar{\Omega})$-overtaking optimal program $\{y_t^*\}_{t=0}^{\infty}$ such that

$$\psi_g(x_0^*, y_0^*) = \sup(\psi_g)$$

and that for all $t = 0, \ldots, T_0$,

$$\rho(x_t, x_t^*) \le \epsilon, \ \rho(y_t, y_t^*) \le \epsilon.$$

Proof Assume that the lemma does not hold. Then there exist a sequence $\{\delta_k\}_{k=1}^{\infty} \subset (0, 1]$, a sequence of (Ω)-programs $\{x_t^{(k)}\}_{t=0}^{T_0}, k = 1, 2, \ldots$ and a sequence of $(\bar{\Omega})$-programs $\{y_t^{(k)}\}_{t=0}^{T_0}, k = 1, 2, \ldots$ such that

$$\lim_{k \to \infty} \delta_k = 0, \tag{5.89}$$

for every integer $k \ge 1$,

$$\psi_g\left(x_0^{(k)}, y_0^{(k)}\right) + \delta_k \ge \sup(\psi_g), \tag{5.90}$$

$$\sum_{t=0}^{T_0-1} \left(u(x_t^{(k)}, x_{t+1}^{(k)}) - u(\bar{x}, \bar{x})\right) - \pi^u\left(x_0^{(k)}\right) + \pi^u\left(x_{T_0}^{(k)}\right) \ge -\delta_k, \tag{5.91}$$

$$\sum_{t=0}^{T_0-1} \left(\bar{u}(y_t^{(k)}, y_{t+1}^{(k)}) - u(\bar{x}, \bar{x})\right) - \pi^{\bar{u}}\left(y_0^{(k)}\right) + \pi^{\bar{u}}\left(y_{T_0}^{(k)}\right) \ge -\delta_k \tag{5.92}$$

and that the following property holds:

(i) for each (u, Ω)-overtaking optimal program $\{z_t\}_{t=0}^{\infty}$ and each $(\bar{u}, \bar{\Omega})$-overtaking optimal program $\{\xi_t\}_{t=0}^{\infty}$ satisfying

$$\psi_g(z_0, \xi_0) = \sup(\psi_g),$$

we have

$$\max\left\{\rho\left(z_t, x_t^{(k)}\right) + \rho\left(\xi_t, y_t^{(k)}\right) : t = 0, \ldots, T_0\right\} > \epsilon.$$

Extracting a subsequence and re-indexing, if necessary, we may assume without loss of generality that for every integer $t \in [0, T_0]$ there exist

$$x_t = \lim_{k \to \infty} x_t^{(k)}, \ y_t = \lim_{k \to \infty} y_t^{(k)}. \tag{5.93}$$

It follows from (5.88)–(5.90) and (5.93) and upper semicontinuity of the functions g, π^u, and $\pi^{\bar{u}}$ that

$$\pi^u(x_0) + \pi^{\bar{u}}(y_0) + g(x_0, y_0) \geq \pi^u(\xi) + \pi^{\bar{u}}(\eta) + g(\xi, \eta) \text{ for all } \xi, \eta \in K,$$
(5.94)

$$\pi^u(x_0) \geq \limsup_{k \to \infty} \pi^u\left(x_0^{(k)}\right),$$
(5.95)

$$\pi^{\bar{u}}(y_0) \geq \limsup_{k \to \infty} \pi^{\bar{u}}\left(y_0^{(k)}\right),$$
(5.96)

$$g(x_0, y_0) \geq \limsup_{k \to \infty} g\left(x_0^{(k)}, y_0^{(k)}\right).$$
(5.97)

Relations (5.89), (5.90), and (5.94) imply that

$$\pi^u(x_0) + \pi^{\bar{u}}(y_0) + g(x_0, y_0)$$
$$= \lim_{k \to \infty} \left(\pi^u\left(x_0^{(k)}\right) + \pi^{\bar{u}}\left(y_0^{(k)}\right) + g\left(x_0^{(k)}, y_0^{(k)}\right)\right).$$
(5.98)

By (5.95)–(5.98),

$$\pi^u(x_0) = \lim_{k \to \infty} \pi^u\left(x_0^{(k)}\right),$$
(5.99)

$$\pi^{\bar{u}}(y_0) = \lim_{k \to \infty} \pi^{\bar{u}}\left(y_0^{(k)}\right),$$
(5.100)

$$g(x_0, y_0) = \lim_{k \to \infty} g\left(x_0^{(k)}, y_0^{(k)}\right).$$
(5.101)

By upper semicontinuity of the functions u, π^u, and $\pi^{\bar{u}}$, (5.80), (5.91)–(5.93), (5.99), and (5.100), we have

$$\sum_{t=0}^{T_0-1} (u(x_t, x_{t+1}) - u(\bar{x}, \bar{x})) - \pi^u(x_0) + \pi^u(x_{T_0})$$

$$\geq \limsup_{k \to \infty} \left(\sum_{t=0}^{T_0-1} \left(u(x_t^{(k)}, x_{t+1}^{(k)}) - u(\bar{x}, \bar{x})\right) - \pi^u\left(x_0^{(k)}\right) + \pi^u\left(x_{T_0}^{(k)}\right)\right)$$

$$\geq \limsup_{k \to \infty} (-\delta_k) = 0,$$

$$\sum_{t=0}^{T_0-1} (\bar{u}(y_t, y_{t+1}) - u(\bar{x}, \bar{x})) - \pi^{\bar{u}}(y_0) + \pi^{\bar{u}}(y_{T_0})$$

$$\geq \limsup_{k\to\infty} \left(\sum_{t=0}^{T_0-1} \left(u\left(y_t^{(k)}, y_{t+1}^{(k)}\right) - u(\bar{x}, \bar{x}) \right) - \pi^{\bar{u}}\left(y_0^{(k)}\right) + \pi^{\bar{u}}\left(y_{T_0}^{(k)}\right) \right)$$

$$\geq \limsup_{k\to\infty}(-\delta_k) = 0.$$

Together with Proposition 5.1 this implies that for all integers $t = 0, \ldots, T_0 - 1$, we have

$$u(x_t, x_{t+1}) - u(\bar{x}, \bar{x}) = \pi^u(x_t) - \pi^u(x_{t+1}), \tag{5.102}$$

$$\bar{u}(y_t, y_{t+1}) - u(\bar{x}, \bar{x}) = \pi^{\bar{u}}(y_t) - \pi^{\bar{u}}(y_{t+1}). \tag{5.103}$$

It follows from Theorem 3.4 that there exist a (u, Ω)-overtaking optimal program $\{\tilde{x}_t\}_{t=0}^{\infty}$ satisfying

$$\tilde{x}_0 = x_{T_0} \tag{5.104}$$

and a $(\bar{u}, \bar{\Omega})$-overtaking optimal program $\{\tilde{y}_t\}_{t=0}^{\infty}$ satisfying

$$\tilde{y}_0 = y_{T_0}. \tag{5.105}$$

For all integers $t > T_0$ put

$$x_t = \tilde{x}_{t-T_0}, \; y_t = \tilde{y}_{t-T_0}. \tag{5.106}$$

Clearly, $\{x_t\}_{t=0}^{\infty}$ is an (Ω)-program and $\{y_t\}_{t=0}^{\infty}$ is an $(\bar{\Omega})$-program. Corollary 5.3 implies that (5.102) and (5.103) hold for all integers $t \geq 0$. It follows from (5.102), (5.103), and Proposition 5.8 that $\{x_t\}_{t=0}^{\infty}$ is a (u, Ω)-overtaking optimal program and $\{y_t\}_{t=0}^{\infty}$ is a $(\bar{u}, \bar{\Omega})$-overtaking optimal program satisfying (5.98). In view of (5.93), for all sufficiently large natural numbers k,

$$\rho(x_t, x_t^{(k)}) \leq \epsilon/4, \; \rho(y_t, y_t^{(k)}) \leq \epsilon/4, \; t = 0, \ldots, T_0.$$

Together with (5.98) this contradicts property (i). The contradiction we have reached proves Lemma 5.18.

5.10   Proof of Theorem 5.17

Lemma 5.18 implies that there exists a number $\delta_1 \in (0, \epsilon/2)$ such that the following property holds:

(i) for every (Ω)-program $\{\xi_t\}_{t=0}^{\tau_0}$ and every $(\bar{\Omega})$-program $\{\eta_t\}_{t=0}^{\tau_0}$ which satisfy

$$\pi^u(\xi_0)+\pi^{\bar{u}}(\eta_0)+g(\xi_0,\eta_0)+2\delta_1 \geq \pi^u(\xi)+\pi^{\bar{u}}(\eta)+g(\xi,\eta) \text{ for all } \xi,\eta \in K,$$

$$\sum_{t=0}^{\tau_0-1}(u(\xi_t,\xi_{t+1}) - u(\bar{x},\bar{x})) - \pi^u(\xi_0) + \pi^u(\xi_{\tau_0}) \geq -2\delta_1,$$

$$\sum_{t=0}^{\tau_0-1}(\bar{u}(\eta_t,\eta_{t+1}) - u(\bar{x},\bar{x})) - \pi^{\bar{u}}(\eta_0) + \pi^{\bar{u}}(\eta_{\tau_0}) \geq -2\delta_1$$

there exists a (u,Ω)-overtaking optimal program $\{x_t^*\}_{t=0}^{\infty}$ and a $(\bar{u},\bar{\Omega})$-overtaking optimal program $\{y_t^*\}_{t=0}^{\infty}$ such that

$$\psi_g(x_0^*,y_0^*) = \sup(\psi_g)$$

and that for all $t = 0,\ldots,\tau_0$,

$$\rho(\xi_t,x_t^*) \leq \epsilon, \ \rho(\eta_t,y_t^*) \leq \epsilon.$$

Propositions 5.4, 5.5, and assumption (B2) imply that there exists a number $\delta_2 \in (0,\delta_1)$ such that the following properties hold:

(ii) for every point $z \in X$ satisfying $\rho(z,\bar{x}) \leq 2\delta_2$, we have

$$|\pi^u(z)| = |\pi^u(z) - \pi^u(\bar{x})| \leq \delta_1/16,$$

$$|\pi^{\bar{u}}(z)| = |\pi^{\bar{u}}(z) - \pi^{\bar{u}}(\bar{x})| \leq \delta_1/16;$$

(iii) for every $(x,y) \in \Omega$ satisfying

$$\rho(x,\bar{x}) \leq 2\delta_2, \ \rho(y,\bar{x}) \leq 2\delta_2,$$

we have

$$|u(x,y) - u(\bar{x},\bar{x})| \leq (\delta_1/16)(L_* + 1)^{-1}$$

By (B3), there exists $\delta_3 \in (0,\delta_2)$ such that the following property holds:

(iv) for each $z_1, z_2 \in K$ satisfying

$$\rho(z_i,\bar{x}) \leq \delta_3, \ i = 1,2$$

there exists an (Ω)-program $\{\xi_t\}_{t=0}^{L_*}$ such that

$$\xi_0 = z_1, \ \xi_{L_*} = z_2$$

and

$$\rho(\xi_t,\bar{x}) \leq \delta_2, \ t = 0,\ldots,L_*.$$

Theorem 3.12 implies that there exist a natural number L_1 and a positive number $\delta_4 \in (0, \delta_3)$ such that the following property holds:

(v) for every natural number $T > 2L_1$, every finite sequence of bounded functions $u_t : \Omega \to R^1, t = 0, \ldots T - 1$ satisfying

$$\|u_t - u\| \le \delta_3, \ t = 0 \ldots, T - 1$$

and every (Ω)-program $\{x_t\}_{t=0}^T$ which satisfies

$$\sum_{t=0}^{T-1} u_t(x_t, x_{t+1}) \ge U\left(\{u_t\}_{t=0}^{T-1}, \Omega, x_0, x_T\right) - \delta_4$$

the inequality $\rho(x_t, \bar{x}) \le \delta_3$ holds for all $t = L_1, \ldots, T - L_1$.

By Theorem 3.4, there exists a (u, Ω)-overtaking optimal program $\{\xi_t\}_{t=0}^\infty$ and a $(\bar{u}, \bar{\Omega})$-overtaking optimal program $\{\eta_t\}_{t=0}^\infty$ such that

$$\psi_g(\xi_0, \eta_0) = \sup(\psi_g). \tag{5.107}$$

In view of property (v),

$$\rho(\xi_t, \bar{x}) \le \delta_3 \text{ for all integers } t \ge L_1, \tag{5.108}$$

$$\rho(\eta_t, \bar{x}) \le \delta_3 \text{ for all integers } t \ge L_1. \tag{5.109}$$

Choose a positive number δ and an integer $T_0 \ge 1$ such that

$$\delta < (64(2L_1 + L_* + \tau_0 + 8))^{-1} \min\{\delta_1, \delta_2, \delta_3, \delta_4\}, \tag{5.110}$$

$$T_0 > 4L_* + 8L_1 + 4\tau_0 + 16. \tag{5.111}$$

Assume that an integer $T \ge T_0$, a function $h \in \mathcal{M}(K \times K)$ satisfies

$$\|h - g\| \le \delta, \tag{5.112}$$

a finite sequence of bounded functions $u_t : \Omega \to R^1, t = 0, \ldots T - 1$ satisfies

$$\|u_t - u\| \le \delta, \ t = 0 \ldots, T - 1 \tag{5.113}$$

and that $\{x_t\}_{t=0}^T$ is an (Ω)-program which satisfies

$$h(x_0, x_T) + \sum_{t=0}^{T-1} u_t(x_t, x_{t+1}) \ge U\left(h, \{u_t\}_{t=0}^{T-1}, \Omega\right) - \delta. \tag{5.114}$$

In view of (5.110), (5.111), (5.113), (5.114), and property (v),

$$\rho(x_t, \bar{x}) \le \delta_3, \ t = L_1, \ldots, T - L_1. \tag{5.115}$$

Property (iv), (5.108), (5.109), (5.111), and (5.115) imply that there exists an (Ω)-program $\{\tilde{x}_t\}_{t=0}^{T}$ such that

$$\tilde{x}_t = \xi_t, \ t = 0, \ldots, 2L_1 + \tau_0 + 3, \tag{5.116}$$

$$\rho(\tilde{x}_t, \bar{x}) \le \delta_2, \ t = 2L_1 + \tau_0 + 3, \ldots, 2L_1 + L_* + \tau_0 + 3, \tag{5.117}$$

$$\tilde{x}_t = x_t, \ t = 2L_1 + \tau_0 + 3 + L_*, \ldots, T - 2L_1 - \tau_0 - 3 - L_*, \tag{5.118}$$

$$\rho(\tilde{x}_t, \bar{x}) \le \delta_2, \ t = T - 2L_1 - \tau_0 - 3 - L_*, \ldots, T - 2L_1 - \tau_0 - 3, \tag{5.119}$$

$$\tilde{x}_t = \eta_{T-t}, \ t = T - 2L_1 - \tau_0 - 3, \ldots, T. \tag{5.120}$$

It follows from (5.110), (5.112)–(5.114), (5.116)–(5.120), and property (iii) that

$$\delta \ge U(h, \{u_t\}_{t=0}^{T-1}, \Omega) - h(x_0, x_T) - \sum_{t=0}^{T-1} u_t(x_t, x_{t+1})$$

$$\ge h(\tilde{x}_0, \tilde{x}_T) + \sum_{t=0}^{T-1} u_t(\tilde{x}_t, \tilde{x}_{t+1}) - h(x_0, x_T) - \sum_{t=0}^{T-1} u_t(x_t, x_{t+1})$$

$$= h(\xi_0, \eta_0) - h(x_0, x_T) + \sum_{t=0}^{2L_1+\tau_0+2+L_*} (u_t(\tilde{x}_t, \tilde{x}_{t+1}) - u_t(x_t, x_{t+1}))$$

$$+ \sum_{t=T-2L_1-\tau_0-3-L_*}^{T-1} (u_t(\tilde{x}_t, \tilde{x}_{t+1}) - u_t(x_t, x_{t+1}))$$

$$\ge g(\xi_0, \eta_0) - g(x_0, x_T)$$

$$+ \sum_{t=0}^{2L_1+\tau_0+2+L_*} (u(\tilde{x}_t, \tilde{x}_{t+1}) - u(x_t, x_{t+1}))$$

$$+ \sum_{t=T-2L_1-\tau_0-3-L_*}^{T-1} (u(\tilde{x}_t, \tilde{x}_{t+1}) - u(x_t, x_{t+1})) - 2\delta(1 + 2L_1 + \tau_0 + 4 + L_*)$$

$$\ge g(\xi_0, \eta_0) - g(x_0, x_T)$$

$$+ \sum_{t=0}^{2L_1+\tau_0+2} u(\xi_t, \xi_{t+1}) + L_*(u(\bar{x}, \bar{x}) - (\delta_1/16)(L_* + 1)^{-1})$$

$$
-\sum_{t=0}^{2L_1+\tau_0+2+L_*} u_t(x_t, x_{t+1}) + \sum_{t=T-2L_1-\tau_0-3}^{T-1} u(\eta_{T-t}, \eta_{T-t-1})
$$

$$
+ L_*(u(\bar{x}, \bar{x}) - (\delta_1/16)(L_* + 1)^{-1})
$$

$$
-\sum_{t=T-2L_1-\tau_0-3-L_*}^{T-1} u_t(x_t, x_{t+1}) - \delta_1/16
$$

$$
\geq g(\xi_0, \eta_0) - g(x_0, x_T)
$$

$$
+ \sum_{t=0}^{2L_1+\tau_0+2} u(\xi_t, \xi_{t+1}) + \sum_{t=0}^{2L_1+\tau_0+2} \bar{u}(\eta_t, \eta_{t+1}) + 2L_*u(\bar{x}, \bar{x})
$$

$$
-\sum_{t=0}^{2L_1+\tau_0+2+L_*} u(x_t, x_{t+1}) - \sum_{t=T-2L_1-\tau_0-3-L_*}^{T-1} u(x_t, x_{t+1}) - 3\delta_1/16.
$$

$$(5.121)$$

Define

$$
y_t = x_{T-t}, \quad t = 0, \dots, T.
$$

$$(5.122)$$

Relations (5.22), (5.121), and (5.122) imply that

$$
g(x_0, y_0) + \sum_{t=0}^{2L_1+\tau_0+L_*+2} u(x_t, x_{t+1}) + \sum_{t=0}^{2L_1+\tau_0+L_*+2} \bar{u}(y_t, y_{t+1})
$$

$$
\geq g(\xi_0, \eta_0) + \sum_{t=0}^{2L_1+\tau_0+2} u(\xi_t, \xi_{t+1}) + \sum_{t=0}^{2L_1+\tau_0+2} \bar{u}(\eta_t, \eta_{t+1})
$$

$$
+ 2L_*u(\bar{x}, \bar{x}) - 3\delta_1/16 - \delta_1/16.
$$

$$(5.123)$$

It follows from (5.108), (5.109), (5.111), (5.112), (5.115), (5.123), Proposition 5.1, Corollary 5.3, and properties (ii) and (iii) that

$$
g(x_0, y_0) - g(\xi_0, \eta_0)
$$

$$
+ \sum_{t=0}^{\tau_0-1} (u(x_t, x_{t+1}) - u(\bar{x}, \bar{x})) - \pi^u(x_0) + \pi^u(x_{\tau_0})
$$

$$
+ \sum_{t=0}^{\tau_0-1} (\bar{u}(y_t, y_{t+1}) - u(\bar{x}, \bar{x})) - \pi^{\bar{u}}(y_0) + \pi^{\bar{u}}(y_{\tau_0})
$$

$$\geq g(x_0, y_0) - g(\xi_0, \eta_0)$$

$$+ \sum_{t=0}^{2L_1+\tau_0+2+L_*} (u(x_t, x_{t+1}) - u(\bar{x}, \bar{x})) - \pi^u(x_0) + \pi^u(x_{2L_1+\tau_0+L_*+3})$$

$$+ \sum_{t=0}^{2L_1+\tau_0+L_*+2} (\bar{u}(y_t, y_{t+1}) - \bar{u}(\bar{x}, \bar{x})) - \pi^{\bar{u}}(y_0) + \pi^{\bar{u}}(y_{2L_1+\tau_0+3+L_*})$$

$$\geq \sum_{t=0}^{2L_1+\tau_0+2} (u(\xi_t, \xi_{t+1}) - u(\bar{x}, \bar{x})) - \pi^u(x_0) + \pi^u(x_{2L_1+\tau_0+3+L_*})$$

$$+ \sum_{t=0}^{2L_1+\tau_0+2} (\bar{u}(\eta_t, \eta_{t+1}) - \bar{u}(\bar{x}, \bar{x})) - \pi^{\bar{u}}(y_0) + \pi^{\bar{u}}(y_{2L_1+\tau_0+3+L_*}) - \delta_1/4$$

$$\geq \pi^u(\xi_0) - \pi^u(\xi_{2L_1+\tau_0+3}) - \pi^u(x_0) + \pi^u(x_{2L_1+\tau_0+3+L_*})$$

$$+ \pi^{\bar{u}}(\eta_0) - \pi^{\bar{u}}(\eta_{2L_1+\tau_0+3})$$

$$- \pi^{\bar{u}}(y_0) - \pi^{\bar{u}}(y_{2L_1+\tau_0+3+L_*}) - \delta_1/4$$

$$\geq \pi^u(\xi_0) - \pi^u(x_0) + \pi^{\bar{u}}(\eta_0) - \pi^{\bar{u}}(y_0) - \delta_1/4 - \delta_1/4. \tag{5.124}$$

In view of (5.124), we have

$$g(x_0, y_0) + \pi^u(x_0) + \pi^{\bar{u}}(y_0) - (g(\xi_0, \eta_0) + \pi^u(\xi_0) + \pi^{\bar{u}}(\eta_0))$$

$$+ \sum_{t=0}^{\tau_0-1} (u(x_t, x_{t+1}) - u(\bar{x}, \bar{x})) - \pi^u(x_0) + \pi^u(x_{\tau_0})$$

$$+ \sum_{t=0}^{\tau_0-1} (\bar{u}(y_t, y_{t+1}) - u(\bar{x}, \bar{x})) - \pi^{\bar{u}}(y_0) + \pi^{\bar{u}}(y_{\tau_0}) \geq -\delta_1. \tag{5.125}$$

By (5.107), (5.125), and Proposition 5.1,

$$g(x_0, y_0) + \pi^u(x_0) + \pi^{\bar{u}}(y_0) \geq \sup(\psi_g) - \delta_1,$$

$$\sum_{t=0}^{\tau_0-1} (u(x_t, x_{t+1}) - u(\bar{x}, \bar{x})) - \pi^u(x_0) + \pi^u(x_{\tau_0}) \geq -\delta_1,$$

$$\sum_{t=0}^{\tau_0-1} (\bar{u}(y_t, y_{t+1}) - u(\bar{x}, \bar{x})) - \pi^{\bar{u}}(y_0) + \pi^{\bar{u}}(y_{\tau_0}) \geq -\delta_1.$$

It follows from the relations above and property (i) that there exists a (u, Ω)-overtaking optimal program $\{x_t^*\}_{t=0}^{\infty}$ and a $(\bar{u}, \bar{\Omega})$-overtaking optimal program $\{y_t^*\}_{t=0}^{\infty}$ such that

$$\psi_g(x_0^*, y_0^*) = \sup(\psi_g)$$

and for all $t = 0, \ldots, \tau_0$, $\rho(x_t, x_t^*) \leq \epsilon$, $\rho(x_{T-t}, y_t^*) = \rho(y_t, y_t^*) \leq \epsilon$. Theorem 5.17 is proved.

Chapter 6
Applications to the Forest Management Problem

In this chapter we continue the discussion of the forest management problem and show that for this problem the results of Chaps. 3 and 5 hold.

6.1 Preliminaries

We consider a discrete time model for the optimal management of a forest of total area S occupied by k species $I = \{1, \ldots, k\}$ with maturity ages of $n_1, \ldots, n_k$ years respectively. This model was introduced in Sect. 2.1. For the reader's convenience we recall here the notation and definitions.

For each period $t = 0, 1, \ldots$ we denote $x_i^j(t) \geq 0$ the area covered by trees of species i that are j years old with $j = 1, \ldots, n_i$ and $\bar{x}_i(t) \geq 0$ the area occupied by over-mature trees (older than n_i).

Assuming that only mature trees can be harvested we must have

$$u_i(t) \leq \bar{x}_i(t) + x_i^{n_i}(t), \tag{6.1}$$

and then the area not harvested in that period will comprise the over-mature trees at the next step, namely

$$\bar{x}_i(t+1) = \bar{x}_i(t) + x_i^{n_i}(t) - u_i(t). \tag{6.2}$$

The fact that immature trees cannot be harvested is represented by

$$x_i^{j+1}(t+1) = x_i^j(t), \quad j = 1, \ldots, n_i - 1. \tag{6.3}$$

© The Author(s), under exclusive license to Springer Nature Switzerland AG 2019
A. J. Zaslavski, *Optimal Control Problems Arising in Forest Management*,
SpringerBriefs in Optimization, https://doi.org/10.1007/978-3-030-23587-1_6

The total harvested area $\sum_{i \in I} u_i(t)$ is allocated to new seedlings which is expressed by the equation

$$\sum_{i \in I} x_i^1(t+1) = \sum_{i \in I} u_i(t). \tag{6.4}$$

In the sequel we use the notation

$$x_i^{n_i+1} = \bar{x}_i, \ i \in I. \tag{6.5}$$

A representation of the forest in terms of the age distribution at time t is provided by the state

$$x(t) = (x_1(t), \ldots, x_k(t)),$$

where $x_i(t) = (x_i^1(t), \ldots, x_i^{n_i}(t), x_i^{n_i+1}(t))$ describes the areas occupied in year t by trees of species i with ages $1, 2, \ldots, n_i$ and over n_i. The first and last components of each vector $x_i(t)$ are controlled by the sowing and harvesting policies.

Let $R_+^m = \{x = (x_1, \ldots, x_m) \in R^m : x_i \geq 0, \ i = 1, \ldots, m\}$.

Let $N = \sum_{i \in I}(n_i + 1)$. Every vector $x \in R^N$ is represented as $x = (x_1, \ldots, x_k)$, where $x_i = (x_i^1, \ldots, x_i^{n_i}, x_i^{n_i+1}) \in R^{n_i+1}$ for all integers $i = 1, \ldots, k$.

Denote by Δ the set of all $x \in R_+^N$ such that

$$\sum_{i \in I} \left[\sum_{j=1}^{n_i+1} x_i^j \right] = S. \tag{6.6}$$

A sequence $\{x(t)\}_{t=0}^{\infty} \subset \Delta$ is called a program if for all integers $t \geq 0$ and all $i \in I$ (6.1)–(6.4) hold (see (6.5)) with some $u(t) = (u_1(t), \ldots, u_k(t)) \in R_+^k$.

Let the integers T_1, T_2 satisfy $0 \leq T_1 < T_2$. A sequence $\{x(t)\}_{t=T_1}^{T_2} \subset \Delta$ is called a program if (6.1)–(6.4) hold for all $i \in I$ and for all integers $t = T_1, \ldots, T_2 - 1$ (see (6.5)) with some $u(t) = (u_1(t), \ldots, u_k(t)) \in R_+^k$.

An alternative equivalent definition of a program is given with the help of the transition possibility. Define

$$\Omega = \Big\{(x, y) \in \Delta \times \Delta : \ y_i^{j+1} = x_i^j \text{ for all } i \in I \text{ and all } j \in \{1, \ldots, n_i\} \setminus \{n_i\}$$

$$\text{and for all } i \in I, \ x_i^{n_i+1} + x_i^{n_i} - y_i^{n_i+1} \geq 0\Big\}. \tag{6.7}$$

Evidently, if $(x, y) \in \Omega$, then

$$\sum_{i \in I} y_i^1 = \sum_{i \in I} \left(x_i^{n(i)+1} + x_i^{n_i} - y_i^{n(i)+1} \right). \tag{6.8}$$

It is not difficult to see that a sequence $\{x(t)\}_{t=0}^{\infty} \subset \Delta$ is a program if and only if $(x(t), x(t+1)) \in \Omega$ for all integers $t \geq 0$.

Let the integers T_1, T_2 satisfy $0 \leq T_1 < T_2$. It is not difficult to see that a sequence $\{x(t)\}_{t=T_1}^{T_2} \subset \Delta$ is a program if and only if $(x(t), x(t+1)) \in \Omega$ for all $t = T_1, \ldots, T_2 - 1$.

For each $(x, y) \in \Omega$ set

$$V(x, y) = (v_1(x, y), \ldots, v_k(x, y)),$$

where for $i = 1, \ldots, k$,

$$v_i(x, y) = x_i^{n_i+1} + x_i^{n_i} - y_i^{n_i+1}.$$

Set

$$\Delta_0 = \left\{ x \in R_+^k : \sum_{i=1}^{k} x_i \leq S \right\}.$$

In this chapter we assume that a benefit at moment $t = 0, 1, \ldots$ is represented by an upper semicontinuous function $w_t : \Delta_0 \to R^1$ and at a moment $t = 0, 1, \ldots$, $w_t(V(x, y))$ is the benefit obtained today if the forest today is x and the forest tomorrow is y, where $(x, y) \in \Omega$.

We suppose that $w_t = w$ for all integers $t \geq 0$.

Evidently, Δ is a compact set in R^N, Ω is a closed subset of $\Delta \times \Delta$ and $w \circ V : \Omega \to R^1$ is an upper semicontinuous function. Put

$$\bar{n} = \max\{n_i : i \in I\}. \tag{6.9}$$

This model is a particular case of the general optimal control system studied in Chaps. 2, 3, and 5. Proposition 2.1 implies that assumption (A) holds with

$$\bar{L} = N + \bar{n} + 1.$$

All the results of Chap. 2 hold for the particular case considered here with $u = w \circ V$.

We use the notation and definitions introduced and used in Chaps. 2 and 3.

Let $y, z \in \Delta$ and let the integers T_1, T_2 satisfy $T_1 < T_2$. Set

$$U(y, T_1, T_2) = \sup \left\{ \sum_{t=T_1}^{T_2-1} w(V(x_t, x_{t+1})) : \right.$$

$$\left. \{x_t\}_{t=T_1}^{T_2} \text{ is a program and } x_{T_1} = y \right\},$$

$$U(y, \tilde{y}, T_1, T_2) = \sup \left\{ \sum_{t=T_1}^{T_2-1} w(V(x_t, x_{t_{t+1}})) : \right.$$

$$\left. \{x_t\}_{t=T_1}^{T_2} \text{ is a program and } x_{T_1} = y, \ x_{T_2} = \tilde{y} \right\},$$

$$\widehat{U}(T_1, T_2) = \sup \left\{ \sum_{t=T_1}^{T_2-1} w(V(x_t, x_{t_{t+1}})) : \{x_t\}_{t=T_1}^{T_2} \text{ is a program} \right\}.$$

Let

$$\mu := \mu(w \circ V)$$

be defined by Theorem 2.9.

For any $x = (x_1, \ldots, x_m) \in R^m$ set

$$\|x\| = \max\{|x_i| : \ i = 1, \ldots, m\}.$$

6.2 Auxiliary Results

Lemma 6.1 *Assume that*

$$w(x) \le w(y) \tag{6.10}$$

for each

$$x = (x_1, \ldots, x_k), \ y = (y_1, \ldots, y_k) \in \Delta_0$$

satisfying

$$x_i \le y_i, \ i = 1, \ldots, k$$

and that

$$\mu(w \circ V) = \sup\{(w \circ V)(x, x) : \ x \in \Delta, \ (x.x) \in \Omega\}. \tag{6.11}$$

Then there exists

$$\widehat{x} = (\widehat{x}_1, \ldots, \widehat{x}_k) \in \Delta \tag{6.12}$$

such that

$$\widehat{x}_i = \left(\widehat{x}_i^1 \ldots, \widehat{x}_i^{n_i+1} \right), \ i = 1, \ldots, k,$$

$$(\widehat{x}, \widehat{x}) \in \Omega, \tag{6.13}$$

$$w(V(\widehat{x}, \widehat{x})) \geq w(V(x, x)) \text{ for all } x \in \Delta \text{ such that } (x, x) \in \Omega, \tag{6.14}$$

and that for each $i = 1, \ldots, k,$

$$\widehat{x}_i^j = \widehat{x}_i^1, \ j = 1, \ldots, n_i, \ \widehat{x}_i^{n_i+1} = 0.$$

Proof In view of (6.11), there exists $\widehat{x} = (\widehat{x}_1, \ldots, \widehat{x}_k) \in \Delta$ satisfying (6.12)–(6.14). By (6.12), (6.13), and the definition of Ω, for each $i = 1, \ldots, k,$

$$\widehat{x}_i^j = \widehat{x}_i^1, \ j = 1, \ldots, n_i.$$

If

$$\widehat{x}_i^{n_i+1} = 0, \ i = 1, \ldots, k,$$

then the assertion of the lemma holds.

Assume that

$$d := \sum_{i=1}^{k} \widehat{x}_i^{n_i+1} > 0. \tag{6.15}$$

Then

$$V(\widehat{x}, \widehat{x}) = (\widehat{x}_1^1, \ldots, \widehat{x}_k^1)$$

by definition. Define $y = (y_1, \ldots, y_k)$, for $i = 1, \ldots, k$ by

$$y_i^j = \widehat{x}_i^j + d \left(\sum_{p=1}^{k} n_p \right)^{-1}, \ j = 1, \ldots, n_i, \ y_i^{n_i+1} = 0. \tag{6.16}$$

It is not difficult to see that

$$y \in \Delta, \ (y, y) \in \Omega, \ V(y, y) = \left(y_1^1, \ldots, y_k^1 \right). \tag{6.17}$$

It follows from (6.10), (6.11), and (6.17) that

$$w(V(y, y)) \geq w(V(\widehat{x}, \widehat{x})) \geq w(V(x, x))$$

for all $x \in \Delta$ satisfying $(x, x) \in \Omega$. Lemma 6.1 is proved.

Let

$$\widehat{x} = (\widehat{x}_1, \dots, \widehat{x}_k) \in \Delta, \; (\widehat{x}, \widehat{x}) \in \Omega, \tag{6.18}$$

$$\widehat{x}_i = (\widehat{x}_i^1 \dots, \widehat{x}_i^{n_i+1}), \; i = 1, \dots, k,$$

for $i = 1, \dots, k$,

$$\widehat{x}_i^j = \widehat{x}_i^1, \; j = 1, \dots, n_i, \; \widehat{x}_i^{n_i+1} = 0, \tag{6.19}$$

$$\tilde{I} = \{i \in I : \; x_i^1 > 0\}. \tag{6.20}$$

Proposition 6.2 *Let*

$$\epsilon \in (0, \min\{\widehat{x}_i^1 : \; i \in \tilde{I}\}). \tag{6.21}$$

Then for each $x, y \in \Delta$ satisfying

$$\|x - \bar{x}\|, \; \|y - \bar{x}\| \le (2N)^{-1}\epsilon,$$

there is a program $\{x(t)\}_{t=0}^{N+\bar{n}+1}$ such that

$$x(0) = x, \; x(N + \bar{n} + 1) = y,$$

$$\|x(t) - \widehat{x}\| \le \epsilon, \; t = 0, \dots, N + \bar{n} + 1. \tag{6.22}$$

Proof We may assume without loss of generality that

$$\tilde{I} = \{1, \dots, k_0\},$$

where $k_0 \in [1, k]$ is an integer.

Let

$$x, y \in \Delta, \; \|x - \widehat{x}\|, \; \|y - \widehat{x}\| \le \epsilon(2N)^{-1}. \tag{6.23}$$

We construct a program $\{x(t)\}_{t=0}^{N+\bar{n}+1}$ which satisfies (6.22). Set $x(0) = x$.

Assume that $t \in \{0, \dots, N + \bar{n}\}$ is an integer and that a program $\{x(p)\}_{p=0}^t$ was defined. In order to define $x(t + 1)$ we need to choose $v(t) \in R_+^k$ and

$$(x_1^1(t + 1), \dots, x_k^1(t + 1)) \in R_+^k$$

such that for all integers $i = 1, \dots, k$ we have

$$v_i(t) \le x_i^{n_i}(t) + x_i^{n_i+1}(t), \; \sum_{i=1}^k x_i^1(t + 1) = \sum_{i=1}^k v_i(t) \tag{6.24}$$

and then the state $x(t + 1)$ is defined as follows for all integers $i = 1, \ldots, k$:

$$x_i^j(t + 1) = x_i^{j-1}(t), \quad j \text{ is an integer satisfying } 2 \leq j \leq n_i, \tag{6.25}$$

$$x_i^{n_i+1}(t + 1) = x_i^{n_i}(t) + x_i^{n_i+1}(t) - v_i(t). \tag{6.26}$$

Thus the construction of the program is done as follows: if the states are defined till moment $t \leq N + \bar{n}$, then we choose $V(t) \in R_+^k$ and $(x_1^1(t+1), \ldots, x_k^1(t+1)) \in R_+^k$ such that (6.24) holds and define $x(t + 1)$ by (6.25) and (6.26).

For each $t = 0, \ldots, N - 1$ set

$$x_i^1(t + 1) = v_i(t) = \widehat{x}_i^1 - \epsilon(2N)^{-1}, \quad i = 1, \ldots, k_0,$$

$$x_i^1(t + 1) = v_i(t) = 0 \text{ for all integers } i \text{ satisfying } k_0 < i \leq k \tag{6.27}$$

and define $x_i^j(t + 1)$, $i = 1, \ldots, k$, $j = 2, \ldots, n_i + 1$ by (6.25) and (6.26).

In order to show that $\{x(t)\}_{t=0}^N$ is a program it is sufficient to show that the first part of (6.24) is true for all integers $i = 1, \ldots, k_0$.

It follows from (6.23), (6.27), and (6.25) that for all integers $t = 0, \ldots, N$, all integers $i = 1, \ldots, k_0$, and all integers $j = 1, \ldots, n_i$ we have

$$\widehat{x}_i^1 - \epsilon(2N)^{-1} \leq x_i^j(t) \leq \widehat{x}_i^1 + \epsilon(2N)^{-1}. \tag{6.28}$$

Equations (6.27) and (6.28) imply (6.24) for all integers $i = 1, \ldots, k_0$. Then $\{x(t)\}_{t=0}^N$ is a program. In view of (6.23), (6.27), and (6.28), for all integers $t = 0, \ldots, N$,

$$0 \leq \sum_{i=1}^k x_i^{n_i+1} \leq S - \sum_{i=1}^{k_0} n_i \widehat{x}_i^1 + N\epsilon(2N)^{-1} \leq 2^{-1}\epsilon. \tag{6.29}$$

By (6.19), (6.28), and (6.29), we have

$$\|x(t) - \widehat{x}\| \leq 2^{-1}\epsilon, \quad t = 0, \ldots, N. \tag{6.30}$$

Let us consider the state $x(N)$. In view of (6.25) and (6.27),

$$x_i^j(N) = \widehat{x}_i^1 - (2N)^{-1}\epsilon, \quad j = 1, \ldots, n(i),$$

$$i = 1, \ldots, k_0, \quad x_i^j(N) = 0, \quad j = 1, \ldots, n_i, \tag{6.31}$$

where an integer i satisfies $k_0 < i \leq k$.

We continue to construct the program.

If at the moment t the state $x(t)$ has already been defined, then we define $V(t) \in R_+^k$ and $(x_1^1(t+1), \ldots, x_k^1(t+1)) \in R_+^k$ such that (6.24) holds and then define $x(t+1)$ by (6.25) and (6.26).

For every integer $s = 1, \ldots, \bar{n}$ set

$$I_s = \{i \in \{1, \ldots, k\} : n_i = s\}. \tag{6.32}$$

(Note that for some integers s we can have $I_s = \emptyset$.)

For all integers $i = 1, \ldots, k$ and all integers $t = N, \ldots, N + \bar{n} - 1$ we define $x_i^{(1)}(t+1) \geq 0$, show that there exists a vector $V(t) \in R_+^k$ which satisfies (6.24) and then define $x_i^j(t+1)$, $i = 1, \ldots, k$ where an integer j satisfies $2 \leq j \leq n_i + 1$.

We begin with the definition of $x_i^1(t+1)$, $i = 1, \ldots, k$, $t = N, \ldots, N + \bar{n} - 1$. Let $i \in \{1, \ldots, k\}$. If $i \in I_1$ ($n_i = 1$), then we set

$$x_i^1(t+N) = \max\left\{\widehat{x}_i^1 - \epsilon(2N)^{-1}, 0\right\} \text{ for all integers } t \text{ satisfying } 1 \leq t < \bar{n}, \tag{6.33}$$

$$x_i^1(\bar{n}+N) = y_i^1 + y_i^2.$$

If $i \in I_s$ ($n_i = s$) with $s > 1$, then we set

$$x_i^1(N+t) = \max\left\{\widehat{x}_i^1 - \epsilon(2N)^{-1}, 0\right\} \text{ for all integers } t$$

$$\text{satisfying } 1 \leq t < \bar{n} + 1 - n_i, \tag{6.34}$$

$$x_i^1(\bar{n}+1-s+N) = y_i^{n_i+1} + y_i^1, \tag{6.35}$$

$$x_i^1(N+t) = y_i^{\bar{n}+2-t} \text{ for all integers } t \text{ satisfying } \bar{n} + 1 - s < t \leq \bar{n}. \tag{6.36}$$

We have defined $x_i^1(t+N)$, $i \in I$, $t = 1, \ldots, \bar{n}$. Then we define $x_i^j(t+1)$ by (6.25) for all $t = N, \ldots, N + \bar{n} - 1$, all integers j satisfying $2 \leq j \leq n_i$, and all $i \in I$. Thus we have defined $x_i^j(t+N)$, $t = 1, \ldots, \bar{n}$, $i \in I$, $j = 1, \ldots, n(i)$.

Set

$$\phi(t) = \sum_{i \in I} \sum_{j=1}^{n_i} x_i^j(N+t) - \sum_{i \in I} x_i^{n_i}(N+t)$$

$$+ \sum_{i \in I} x_i^1(N+t+1), \quad t = 0, \ldots, \bar{n} - 1. \tag{6.37}$$

In view of (6.37), (6.35), and (6.33), for all integers $t = 0, \ldots, \bar{n} - 1$, we have

$$\phi(t) = \sum_{i \in I} \sum_{j=1}^{n_i} x_i^j(N+t+1). \tag{6.38}$$

By (6.33)–(6.38), (6.31), and (6.23),

$$\phi(t) \le \phi(t+1) \text{ for all integers } t \in [0, \bar{n}-1]. \tag{6.39}$$

It follows from (6.38) and (6.33)–(6.36) that

$$\phi(\bar{n}-1) = \sum_{i \in I} \sum_{j=1}^{n_i} x_i^j (N+\bar{n}) = \sum_{i \in I} \sum_{j=1}^{n_i+1} y_i^j = S.$$

Hence

$$\phi(t) \le S, \ t = 0, \ldots, \bar{n}-1. \tag{6.40}$$

In order to complete the construction of the program $\{x(t)\}_{t=0}^{N+\bar{n}}$ we need to determine $V(t) \in R_+^k$ $t = N, \ldots, N + \bar{n} - 1$ and $x_i^{n_i+1}(t+1) \ge 0, \ t = N, \ldots, N + \bar{n} - 1]$ such that the second part of (6.24) is valid for all integers $t = N, \ldots, N + \bar{n} - 1$ and

$$x(t) \in \Delta, \ t = N+1, \ldots, N+\bar{n}. \tag{6.41}$$

We complete this construction by induction. Let $t = 0$. In view of (6.37) and (6.40),

$$\sum_{i \in I} x_i^1 (N+1) \le \sum_{i \in I} \left(x_i^{n_i}(N) + x_i^{n_i+1}(N) \right). \tag{6.42}$$

Relation (6.42) implies that there exists $V(N) \in R_+^k$ such that

$$\sum_{i \in I} x_i^1 (N+1) = \sum_{i \in I} v_i(N), \ v_i(N) \le x_i^{n_i}(N) + x_i^{n_i+1}(N), \ i \in I. \tag{6.43}$$

Define $x_i^{n_i+1}(N+1), i \in I$ by (6.20) with $t = N$.

Assume that an integer τ satisfies $1 \le \tau < \bar{n}$, we have defined $V(t) \in R_+^k$, $t = N, \ldots, N + \tau - 1$, $x_i^{n_i+1}(t), \ i \in I, t = N+1, \ldots, N+\tau$, and $x(t) \in \Delta$, $t = N+1 \ldots, N+\tau$ such that (6.24) is valid for all integers $t = N, \ldots, N+\tau-1$. (Note that for $\tau = 1$ this assumption holds.)

By (6.37) and (6.40),

$$\sum_{i \in I} x_i^1 (N+\tau+1) \le \sum_{i \in I} \left(x_i^{n_i+1}(N+\tau) + x_i^{n_i}(N+\tau) \right).$$

This implies that there is $V(N+\tau) \in R_+^k$ such that

$$\sum_{i \in I} x_i^1 (N+\tau+1) = \sum_{i \in I} V_i(N+\tau),$$

$$V_i(N+\tau) \le x_i^{n_i+1}(N+\tau) + x_i^{n_i}(N+\tau), \ i \in I.$$

Define $x_i^{n_i+1}(N+\tau+1)$ by (6.26) with $t = N+\tau$. Clearly, the assumption made for τ also holds for $\tau + 1$. Thus by induction we defined $x_i^{n_i+1}(t+1) \geq 0$, $t = N, \ldots, N+\bar{n}-1$, $V(t) \in R_+^k$, $t = N, \ldots, N+\bar{n}-1$ such that (6.31) holds and (6.24) holds for all $t = N, \ldots, N+\bar{n}-1$.

Thus we have constructed a program $\{x(t)\}_{t=0}^{N+\bar{n}}$. It follows from (6.33)–(6.36) that for all integers $i \in I$,

$$x_i^{n_i}(N+\bar{n}) = y_i^{n_i+1} + y_i^1,$$

$$x_i^j(N+\bar{n}) = y_i^{j+1} \text{ for all integers } j \text{ satisfying } 1 \leq j < n(i),$$

$$x_i^1(N+\bar{n}) = y_i^2.$$

Evidently,

$$\sum_{i \in I} \sum_{j=1}^{n_i} x_i^j = \sum_{i \in I} \sum_{j=1}^{n_i+1} y_i^j = S.$$

Set $x(N+\bar{n}+1) = y$. It is clear that $(x(N+\bar{n}), y) \in \Omega$. Hence $\{x(t)\}_{t=0}^{N+\bar{n}+1}$ is a program such that $x(0) = x$, $x(N+\bar{n}+1) = y$. In view of (6.33)–(6.36),

$$\|x(t) - \bar{x}\| \leq \epsilon, \ t = 0, \ldots, N+\bar{n}+1.$$

This completes the proof of Proposition 6.2. $\blacksquare$

6.3 Turnpike Results

In the sequel the notation $x \geq y$, $x > y$, $x \gg y$ where $x, y \in R^q$ have their usual meaning. We assume that $w : \Delta_0 \to R^1$ is a continuous and a strictly concave function such that

$$w(x) \leq w(y) \text{ for all } x, y \in \Delta_0 \text{ satisfying } x \leq y. \tag{6.44}$$

Note that $w(V(x, y))$ is the benefit obtained today if the forest today is x and the forest tomorrow is y where $(x, y) \in \Omega$.

In the sequel we assume that supremum over empty set is $-\infty$.

Theorem 2.9 implies the following result.

Theorem 6.3 *There exists $M > 0$ and the limit*

$$\mu := \mu(w \circ V) := \lim_{p \to \infty} \widehat{U}(0, p)/p \tag{6.45}$$

such that

$$|p^{-1}\widehat{U}(0, p) - \mu| \le M/p \text{ for all natural numbers } p.$$

Clearly Ω is a convex set and $w \circ V$ is a concave function. Let μ be as guaranteed by Theorem 6.3. Then Theorem 2.10 implies the following result.

Theorem 6.4

$$\mu = \sup\{w(V(x, x)) : x \in \Delta \text{ and } (x, x) \in \Omega\}.$$

Lemma 6.1 implies that there exists

$$\widehat{x} = (\widehat{x}_1, \dots, \widehat{x}_k) \in \Delta$$

such that

$$(\widehat{x}, \widehat{x}) \in \Omega,$$

for $i = 1, \dots, k$,

$$\widehat{x}_i = \left(\widehat{x}_i^1 \dots, \widehat{x}_i^{n_i+1}\right),$$

$$\widehat{x}_i^j = \widehat{x}_i^1, \ j = 1, \dots, n_i, \ \widehat{x}_i^{n_i+1} = 0,$$

$$\mu = w(V(\widehat{x}, \widehat{x})) \ge w(V(x, x)) \text{ for all } x \in \Delta \text{ satisfying } (x, x) \in \Omega.$$

Proposition 6.2 implies that (B3) holds. It is clear that (B2) holds too.

Theorem 6.5 *Assume that a program $\{x(t)\}_{t=0}^\infty$ is good. Then*

$$\lim_{t \to \infty} x(t) = \widehat{x}.$$

Proof First we show that

$$\lim_{t \to \infty} V(x(t), x(t+1)) = V(\widehat{x}, \widehat{x}). \tag{6.46}$$

For all nonnegative integers t set

$$z(t) = 2^{-1}(x(t) + \widehat{x}). \tag{6.47}$$

Evidently, $\{z(t)\}_{t=0}^\infty$ is a program and

$$V(z(t), z(t+1)) = 2^{-1}V(x(t), x(t+1)) + 2^{-1}V(\widehat{x}, \widehat{x}). \tag{6.48}$$

It follows from (6.48) and the concavity of the function w that $\{z(t)\}_{t=0}^{\infty}$ is a good program and the sequence

$$\left\{\sum_{t=0}^{T-1} w(V(z(t), z(t+1))) - T\mu\right\}_{T=1}^{\infty} \quad \text{is bounded.} \tag{6.49}$$

Assume that (6.46) does not hold. There exist a positive number ϵ and a strictly increasing sequence of natural numbers $\{t_j\}_{j=1}^{\infty}$ such that

$$\|V(\widehat{x}, \widehat{x}) - V(x(t_j), x(t_j+1))\| \geq \epsilon. \tag{6.50}$$

Since the function w is continuous and strictly concave there exists a positive number δ such that for every pair of points $y_1, y_2 \in \Delta_0$ which satisfies $\|y_1 - y_2\| \geq \epsilon$, we have

$$w\left(2^{-1}(y_1 + y_2)\right) - 2^{-1}w(y_1) - 2^{-1}w(y_2) \geq \delta. \tag{6.51}$$

Since the program $\{x_t\}_{t=0}^{\infty}$ is good there exists a positive number M_0 such that for every integer $T > 0$, we have

$$\left|\sum_{t=0}^{T-1}(w(V(x(t), x(t+1))) - \mu)\right| \leq M_0. \tag{6.52}$$

In view of (6.48), the strict concavity of the function w, (6.50) and the choice of the number δ (see (6.51)), (6.52), for every integer $k > 0$,

$$\sum_{t=0}^{t_k+1} w(V(z(t), z(t+1))) - t_k\mu - 2\mu$$

$$= \sum_{t=0}^{t_k+1} \left[w(2^{-1}V(x(t), x(t+1)) + 2^{-1}V(\widehat{x}, \widehat{x}))\right] - t_k\mu - 2\mu$$

$$= \sum_{t=0}^{t_k+1} [w(2^{-1}V(x(t), x(t+1))$$

$$+ 2^{-1}V(\widehat{x}, \widehat{x})) - 2^{-1}w(V(x(t), x(t+1))) - 2^{-1}w(V(\widehat{x}, \widehat{x}))]$$

$$- t_k\mu - 2\mu + \sum_{t=0}^{t_k+1} 2^{-1}w(V(x(t), x(t+1))) + 2^{-1}(t_k+2)w(V(\widehat{x}, \widehat{x}))$$

$$\geq \delta k + 2^{-1}\sum_{t=0}^{t_k+1} [w(V(x(t), x(t+1))) - \mu]$$

$$\geq \delta k - M_0 \to \infty \text{ as } k \to \infty,$$

a contradiction. The contradiction we have reached proves that (6.46) is valid.

Set

$$\widehat{V} = V(\widehat{x}, \widehat{x}) = (\widehat{x}_1^1, \widehat{x}_2^1, \ldots, \widehat{x}_k^1) \tag{6.53}$$

and for all integers $t \geq 0$ set

$$V(t) = V(x(t), x(t+1)), \tag{6.54}$$

where

$$V(t) = (v_1(t), \ldots, v_k(t)). \tag{6.55}$$

Let $t \geq 0$ be an integer. In view of (6.2), (6.5), (6.26), and (6.54), for all integers $i = 1, \ldots, k$,

$$x_i^{n_i}(t) = x_i^{n_i+1}(t+1) - x_i^{n_i+1}(t) + v_i(t),$$
$$x_i^{n_i+1}(t+1) = x_i^{n_i+1}(t) + x_i^{n_i}(t) - v_i(t), \quad i = 1, \ldots, k \tag{6.56}$$

and

$$x_i^{j+1}(t+1) = x_i^j(t), \quad j = 1, \ldots, n_i - 1 \tag{6.57}$$

for all integers $i = 1, \ldots, k$ if $n_i \geq 2$. It follows from (6.56) and (6.57) that for all integers $T \geq 1$ and for all integers $i = 1, \ldots, k$,

$$\sum_{t=T}^{T+n_i-1} v_i(t) = \sum_{t=T}^{T+n_i-1} \left[x_i^{n_i+1}(t) + x_i^{n_i}(t) - x_i^{n_i+1}(t+1) \right]$$

$$= x_i^{n_i+1}(T) - x_i^{n_i+1}(T+n_i) + \sum_{t=T}^{T+n_i-1} x_i^{n_i}(t)$$

$$= x_i^{n_i+1}(T) - x_i^{n_i+1}(T+n_i) + \sum_{j=1}^{n_i} x_i^j(T). \tag{6.58}$$

By (6.58)

$$\sum_{i \in I} \left[\sum_{t=T}^{T+n_i-1} v_i(t) \right] = \sum_{i \in I} \left[x_i^{n_i+1}(T) - x_i^{n+1}(T+n_i) \right] + \sum_{i \in I} \left[\sum_{j=1}^{n_i} x_i^j(T) \right]$$

$$= \sum_{i \in I} \left[\sum_{j=1}^{n_i+1} x_i^j(T) \right] - \sum_{i \in I} x_i^{n_i+1}(T+n_i)$$

$$= S - \sum_{i \in I} x_i^{n_i+1}(T+n_i). \tag{6.59}$$

By (6.46), (6.53)–(6.55), and (6.59), we have

$$\sum_{i \in I} n_i \widehat{x}_i^1 = \lim_{T \to \infty} \sum_{i \in I} \left[\sum_{t=T}^{T+n_i-1} v_i(t) \right] = S - \lim_{T \to \infty} \sum_{i \in I} x_i^{n_i+1}(T + n_i). \qquad (6.60)$$

By (6.53), (6.60), the choice of $\widehat{x}$, we have

$$\lim_{T \to \infty} x_i^{n_i+1}(T + n_i) = 0, \ i = 1, \ldots, k. \qquad (6.61)$$

It follows from (6.56) and (6.61) that for all integers $i = 1, \ldots, k$,

$$\lim_{t \to \infty} (x_i^{n_i}(t) - v_i(t)) = 0. \qquad (6.62)$$

In view of (6.46), (6.59), and (6.62), for all integers $i = 1, \ldots, k$,

$$\lim_{t \to \infty} x_i^{n_i}(t) = \lim_{t \to \infty} v_i(t) = V_i(\widehat{x}, \widehat{x}) = \widehat{x}_i^1. \qquad (6.63)$$

By (6.63), for all integers $i = 1, \ldots, k$ and every integer j which satisfies $1 \leq j \leq n_i$,

$$\lim_{t \to \infty} x_i^j(t) = \lim_{t \to \infty} x_i^{n_i}(t + n_i - j) = \widehat{x}_i^1.$$

Combined with (6.61) this implies that

$$\lim_{t \to \infty} x(t) = \widehat{x}.$$

This completes the proof of Theorem 6.5.

Thus (B1) holds for $w \circ V$ and all the results of Chaps. 3 and 5 are true in our particular case.

6.4 Generic Results

Denote by $C(\Delta_0)$ the set of all continuous functions on the set

$$\Delta_0 = \{v \in R_+^k : \sum_{i=1}^k v_i \leq S\},$$

where $S > 0$. For each $f \in C(\Delta_0)$ set

$$\|f\| = \sup\{|f(x)| : x \in \Delta_0\}.$$

We continue to study the forest management problem which is a particular case of the optimal control problem considered in Chaps. 3 and 5 with $u = w \circ V$, where $w \in C(\Delta_0)$. Recall that for each $(x, y) \in \Omega$,

$$V(x, y) = (v_1(x, y), \ldots, v_k(x, y)), \tag{6.64}$$

where for $i = 1, \ldots, k$,

$$v_i(x, y) = x_i^{n_i+1} + x_i^{n_i} - y_i^{n_i+1}. \tag{6.65}$$

Denote by $C_m(\Delta_0)$ the set of all $w \in C(\Delta_0)$ such that $w : \Delta_0 \to [0, \infty)$,

$$w(x) \le w(y) \text{ for all } x, y \in \Delta_0 \text{ satisfying } x \le y,$$

and that

$$\mu(w \circ V) = \sup\{w(V(x, x)) : x \in \Delta \text{ and } (x, x) \in \Omega\}. \tag{6.66}$$

It is not difficult to see that $C_m(\Delta_0)$ is a closed subset of $C(\Delta_0)$. The space $C_m(\Delta_0)$ is equipped with the metric $d_C(f_1, f_2) = \|f_1 - f_2\|$, $f_1, f_2 \in C_m(\Delta_0)$.
    Let $w \in C_m(\Delta_0)$. Lemma 6.1 implies that there exists

$$\hat{x} = (\hat{x}_1, \ldots, \hat{x}_k) \in \Delta$$

such that

$$(\hat{x}, \hat{x}) \in \Omega, \tag{6.67}$$

$$w(V(\hat{x}, \hat{x})) \ge w(V(x, x)) \text{ for all } x \in \Delta \text{ satisfying } (x, x) \in \Omega, \tag{6.68}$$

for $i = 1, \ldots, k$,

$$\hat{x}_i = (\hat{x}_i^1 \ldots, \hat{x}_i^{n_i+1}),$$

$$\hat{x}_i^j = \hat{x}_i^1, \ j = 1, \ldots, n_i, \ \hat{x}_i^{n_i+1} = 0. \tag{6.69}$$

Assume that

$$\mu(w \circ V) = w(V(\hat{x}, \hat{x})) > 0. \tag{6.70}$$

Let $\gamma \in (0, k^{-1})$ and set

$$I_0 = \{i \in \{1, \ldots, k\} : \hat{x}_i^1 > 0\}. \tag{6.71}$$

For each $i \in I_0$ define a function $\phi_i : [0, \infty) \to [0, \infty)$ by

$$\phi_i(x) = w(V(\widehat{x}, \widehat{x})) \text{ for all } x \in [\widehat{x}_i^1, \infty), \tag{6.72}$$

$$\phi_i(x) = x(\widehat{x}_i^1)^{-1} w(V(\widehat{x}, \widehat{x})) \text{ for all } x \in [0, \widehat{x}_i^1]. \tag{6.73}$$

For every $x = (x_1, \ldots, x_k) \in \Delta_0$ define

$$w_\gamma(x) = w(x) + \gamma \sum_{i \in I_0} \phi_i(x_i). \tag{6.74}$$

Clearly, w is a continuous increasing function.

Lemma 6.6 $w_\gamma \in C_m(\Delta_0)$ and for each $(w_\gamma \circ V)$-good program $\{x(t)\}_{t=0}^\infty$,

$$\lim_{t \to \infty} x(t) = \widehat{x}.$$

Proof Theorem 2.9 and (6.70) imply that there exists $M_0 > 0$ such that for each (Ω)-program $\{x(t)\}_{t=0}^\infty$ and each integer $T \geq 1$,

$$\sum_{t=0}^{T-1} w(V(x(t), x(t+1))) \leq Tw(V(\widehat{x}, \widehat{x})) + M_0. \tag{6.75}$$

Let $\{x(t)\}_{t=0}^\infty$ be a program. By (6.72)–(6.75), for each integer $T > 0$,

$$\sum_{t=0}^{T-1} w_\gamma(V(x(t), x(t+1)))$$

$$= \sum_{t=0}^{T-1} w(V(x(t), x(t+1))) + \gamma \sum_{t=0}^{T-1} \sum_{i \in I_0} \phi_i(v_i(x(t), x(t+1)))$$

$$\leq Tw(V(\widehat{x}, \widehat{x})) + M_0 + \gamma T \sum_{i \in I_0} (w \circ V)(\widehat{x}, \widehat{x})$$

$$= M_0 + Tw(V(\widehat{x}, \widehat{x}))(1 + \gamma \mathrm{Card}(I_0)) = M_0 + (w_\gamma \circ V)(\widehat{x}, \widehat{x})T.$$

This implies that

$$\mu(w_\gamma \circ V) = (w_\gamma \circ V)(\widehat{x}, \widehat{x}), \tag{6.76}$$

and that $w_\gamma \in C_m(\Delta_0)$.

Let $\{x(t)\}_{t=0}^\infty$ be a $(w_\gamma \circ V)$-good program. We show that

$$\lim_{t \to \infty} x(t) = \widehat{x}.$$

In view of (6.76), there exists $M_1 > 0$ such that for each integer $T \geq 1$,

$$\left| \sum_{t=0}^{T-1} w_\gamma(V(x(t), x(t+1))) - T w_\gamma(V(\widehat{x}, \widehat{x})) \right| \leq M_1. \tag{6.77}$$

It follows from (6.72)–(6.75) and (6.77) that for all integers $T \geq 1$,

$$
\begin{aligned}
&- M_1 + T w(V(\widehat{x}, \widehat{x}))(1 + \gamma \mathrm{Card}(I_0)) \\
&= -M_1 + T (w_\gamma \circ V)(\widehat{x}, \widehat{x}) \\
&\leq \sum_{t=0}^{T-1} w_\gamma(V(x(t), x(t+1))) \\
&= \sum_{t=0}^{T-1} w(V(x(t), x(t+1))) + \gamma \sum_{t=0}^{T-1} \sum_{i \in I_0} \phi_i(v_i(x(t), x(t+1))) \\
&\leq T (w \circ V)(\widehat{x}, \widehat{x}) + M_0 + \gamma \sum_{t=0}^{T-1} \sum_{i \in I_0} \phi_i(v_i(x(t), x(t+1))). \tag{6.78}
\end{aligned}
$$

By (6.78), for each integer $T > 0$,

$$
M_0 + M_1 \geq \gamma T \mathrm{Card}(I_0)(w \circ V)(\widehat{x}, \widehat{x}) - \gamma \sum_{t=0}^{T-1} \sum_{i \in I_0} \phi_i(v_i(x(t), x(t+1)))
$$

$$
= \gamma \sum_{t=0}^{T-1} \sum_{i \in I_0} ((w \circ V)(\widehat{x}, \widehat{x}) - \phi_i(v_i(x(t), x(t+1)))). \tag{6.79}
$$

Let $i \in I_0$. It follows from (6.72), (6.73), and (6.79) that for every integer $T > 0$,

$$
\gamma^{-1}(M_0 + M_1) \geq \sum_{t=0}^{T-1} ((w \circ V)(\widehat{x}, \widehat{x}) - \phi_i(v_i(x(t), x(t+1)))). \tag{6.80}
$$

Let $\epsilon > 0$. In view of (6.80),

$$
\mathrm{Card}(\{t \in \{0, 1, \dots\} : (w \circ V)(\widehat{x}, \widehat{x}) - \phi_i(v_i(x(t), x(t+1))) \geq \epsilon\}
$$

$$
\leq (\epsilon \gamma)^{-1}(M_0 + M_1)
$$

and there exists an integer $t_\epsilon > 0$ such that for all integers $t \geq t_\epsilon$,

$$
(w \circ V)(\widehat{x}, \widehat{x}) - \phi_i(v_i(x(t), x(t+1))) < \epsilon.
$$

Since ϵ an arbitrary positive number it follows from (6.72) and (6.73) that

$$\liminf_{t\to\infty} v_i((x(t), x(t+1)) \geq \hat{x}_i^1, \ i \in I_0. \tag{6.81}$$

For every nonnegative integer $t \geq 0$, set

$$V(t) = (v_1(t), \ldots, v_k(t)) = V(x(t), x(t+1)). \tag{6.82}$$

Let $t \geq 0$ be an integer. By (6.65) and (6.82) for all $i = 1, \ldots, k$,

$$x_i^{n_i}(t) = x_i^{n_i+1}(t+1) - x_i^{n_i+1}(t) + v_i(t), \tag{6.83}$$

$$x_i^{n_i+1}(t+1) = x_i^{n_i+1}(t) + x_i^{n_i}(t) - v_i(t), \ i = 1, \ldots, k \tag{6.84}$$

$$x_i^{j+1}(t+1) = x_i^j(t), \ j = 1, \ldots, n_i - 1 \tag{6.85}$$

for $i = 1, \ldots, k$ if $n_i \geq 2$. By (6.84) and (6.85) and all integers $T \geq 1$ for all $i = 1, \ldots, k$,

$$\sum_{t=T}^{T+n_i-1} v_i(t) = \sum_{t=T}^{T+n_i-1} \left[x_i^{n_i+1}(t) + x_i^{n_i}(t) - x_i^{n_i+1}(t+1) \right]$$

$$= x_i^{n_i+1}(T) - x_i^{n_i+1}(T+n_i) + \sum_{t=T}^{T+n_i-1} x_i^{n_i}(t)$$

$$= x_i^{n_i+1}(T) - x_i^{n_i+1}(T+n_i) + \sum_{j=1}^{n_i} x_i^j(T). \tag{6.86}$$

By (6.86), for every integer $T \geq 1$,

$$\sum_{i=1}^k \left[\sum_{t=T}^{T+n_i-1} v_i(t) \right] = \sum_{i=1}^k \left[x_i^{n_i+1}(T) - x_i^{n_i+1}(T+n_i) \right] + \sum_{i=1}^k \left[\sum_{j=1}^{n_i} x_i^j(T) \right]$$

$$= \sum_{i=1}^k \left[\sum_{j=1}^{n_i+1} x_i^j(T) \right] - \sum_{i=1}^k x_i^{n_i+1}(T+n_i)$$

$$= S - \sum_{i=1}^k x_i^{n_i+1}(T+n_i). \tag{6.87}$$

In view of (6.67), (6.69), (6.71), (6.81), (6.82), and (6.87),

$$S = \sum_{i=1}^{k} n_i \widehat{x}_i^1$$

$$\leq \sum_{i=1}^{k} \left[\sum_{t=T}^{T+n_i-1} \liminf_{T\to\infty} v_i(x(t), x(t+1)) \right]$$

$$\leq \liminf_{T\to\infty} \sum_{i=1}^{k} \left[\sum_{t=T}^{T+n_i-1} v_i(t) \right]$$

$$= \liminf_{T\to\infty} \left(S - \sum_{i=1}^{k} x_i^{n_i+1}(T+n_i) \right)$$

$$= S - \limsup_{T\to\infty} \sum_{i=1}^{k} x_i^{n_i+1}(T+n_i). \tag{6.88}$$

It follows from (6.88) that

$$\lim_{T\to\infty} x_i^{n_i+1}(T+n_i) = 0, \ i = 1, \dots, k. \tag{6.89}$$

By (6.83) and (6.89) for all $i = 1, \dots, k$

$$\lim_{t\to\infty} (x_i^{n_i}(t) - v_i(t)) = 0. \tag{6.90}$$

By (6.87) and (6.89),

$$\lim_{T\to\infty} \sum_{i=1}^{k} \sum_{t=T}^{T+n_i-1} v_i(t) = S. \tag{6.91}$$

In view of (6.71) and (6.81), for every $i = 1, \dots, k$,

$$\liminf_{t\to\infty} v_i(t) \geq \widehat{x}_i^1, \tag{6.92}$$

By (6.88) and (6.90)–(6.92),

$$\lim_{t\to\infty} v_i(t) = \widehat{x}_i^1, \ i = 1, \dots, k,$$

$$\lim_{t\to\infty} x_i^{n_i}(t) = \widehat{x}_i^1, \ i = 1, \dots, k,$$

$$\lim_{t\to\infty} x_i^j = \widehat{x}_i^1, \ i = 1, \dots, k, \ j = 1, \dots, n_i.$$

Lemma 6.6 is proved.

Theorem 6.7 *There exists a set $\mathcal{F} \subset C_m(\Delta_0)$ which is a countable intersection of open everywhere dense subsets of $C_m(\Delta_0)$ such that for each $w \in \mathcal{F}$ there exists*

$$x^w = (x_1^w, \ldots, x_k^w) \in \Delta$$

such that

$$(x^w, x^w) \in \Omega,$$

for $i = 1, \ldots, k,$

$$x_i^w = (x_i^{w,1} \ldots, x_i^{w,n_i+1}),$$

$$x_i^{w,j} = x_i^{w,1}, \; j = 1, \ldots, n_i, \; x_i^{w,n_i+1} = 0$$

and that for each $(w \circ V)$-good program $\{x(t)\}_{t=0}^{\infty}$,

$$\lim_{t \to \infty} x(t) = x^w.$$

Proof Let $w \in C_m(\Delta_0)$ be such that

$$\mu(w \circ V) > 0,$$

$\gamma \in (0, k^{-1})$ and $p \geq 1$ be an integer. By Lemmas 6.1 and 6.6 and Theorem 3.12, there exist an open neighborhood $\mathcal{U}(w, \gamma, p)$ of w_γ in $C_m(\Delta_0)$, $\delta(w, \gamma, p) > 0$, a natural number $\tau(w, \gamma, p)$ and

$$x^w = (x_1^w, \ldots, x_k^w) \in \Delta$$

such that the following properties hold:

(a)

$$(x^w, x^w) \in \Omega,$$

for $i = 1, \ldots, k,$

$$x_i^w = (x_i^{w,1} \ldots, x_i^{w,n_i+1}),$$

$$x_i^{w,j} = x_i^{w,1}, \; j = 1, \ldots, n_i, \; x_i^{w,n_i+1} = 0,$$

$$\mu(w \circ V) = w(V(x^w, x^w));$$

(b) for each $(w_\gamma \circ V)$-good program $\{x(t)\}_{t=0}^{\infty}$,

$$\lim_{t \to \infty} x(t) = x^w;$$

(c) for each integer $T > 2\tau(w, \gamma, p)$, each $u_t \in \mathcal{U}(w, \gamma, p), t = 0, \ldots, T - 1$ and each program $\{x(t)\}_{t=0}^{T}$ which satisfies

$$\sum_{t=0}^{T-1} (u_t \circ V)(x_t, x_{t+1}) \geq U\left(\{u_t \circ V\}_{t=0}^{T-1}, x_0, x_T\right) - \delta(w, \gamma, p)$$

we have

$$\|x(t) - x^w\| \leq 1/n, \ t = \tau(w, \gamma, p), \ldots, T - \tau(w, \gamma, p).$$

Define

$$\mathcal{F} = \cap_{q=1}^{\infty} \cup \{\mathcal{U}(w, \gamma, p):$$

$$w \in C_m(\Delta_0), \ \mu(w \circ V) > 0, \ \gamma \in (0, 1/k), \ p \geq q \text{ is an integer}\}. \quad \quad (6.93)$$

Clearly, $\mathcal{F}$ is a countable intersection of open everywhere dense sets in $C_m(\Delta_0)$. Let

$$u \in \mathcal{F}, \ \epsilon \in (0, 1). \quad \quad (6.94)$$

Choose a natural number q such that

$$q > 4\epsilon^{-1}. \quad \quad (6.95)$$

It follows from (6.93) and (6.94) that there exist $w \in C_m(\Delta_0)$, $\gamma \in (0, k^{-1})$, and an integer $p \geq q$ such that

$$\mu(w \circ V) > 0,$$

$$u \in \mathcal{U}(w, \gamma, p). \quad \quad (6.96)$$

Assume that

$$\xi_1, \xi_2 \in \Delta,$$

$$(\xi_i, \xi_i) \in \Omega, \ i = 1, 2,$$

$$(u \circ V)(\xi_i, \xi_i) = \mu(u \circ V), \ i = 1, 2.$$

It is not difficult to see that for each $i = 1, 2$ and each integer $T > 0$,

$$U(u \circ V, \xi_i, \xi_i, 0, T) = T(u \circ V)(\xi_i, \xi_i).$$

Combined with property (c) and (6.96) this implies that

$$\|x^u - \xi_i\| \leq p^{-1} \leq q^{-1} < \epsilon/4, \ i = 1, 2, \tag{6.97}$$

$$\|\xi_1 - \xi_2\| < \epsilon.$$

Since ϵ is an arbitrary positive number we conclude that $\xi_1 = \xi_2$ and

$$\{\xi \in \Delta : (\xi, \xi) \in \Omega, \ u(V(\xi, \xi)) = \mu(u \circ V)\}$$

is a singleton.

Denote its unique element by x^u. In view of (6.97),

$$\|x^u - x^w\| \leq p^{-1} < \epsilon. \tag{6.98}$$

Since ϵ is an arbitrary positive number it follows from property (a) that

$$x^u = (x_1^u, \dots, x_k^u) \in \Delta,$$

$$(x^u, x^u) \in \Omega,$$

for $i = 1, \dots, k$,

$$x_i^u = (x_i^{u,1} \dots, x_i^{u,n_i+1}),$$

$$x_i^{u,j} = x_i^{u,1}, \ j = 1, \dots, n_i, \ x_i^{u,n_i+1} = 0.$$

Assume that an integer $T > 2\tau(w, \gamma, p)$ and that a program $\{x_t\}_{t=0}^T$ satisfies

$$\sum_{t=0}^{T-1} (u \circ V)(x_t, x_{t+1}) \geq U(u \circ V, x_0, x_T, 0, T) - \delta(w, \gamma, p).$$

Together with property (c), (6.96), and (6.98) this implies that for all integers $t = \tau(w, \gamma, p), \dots, T - \tau(w, \gamma, p)$,

$$\|x(t) - x^w\| \leq 1/p$$

and

$$\|x(t) - x^u\| \leq 2/p < \epsilon.$$

Combined with Theorem 4.1 this implies that every $(u \circ V)$-good program converges to x^u. Theorem 6.7 is proved.

It is clear that for every $w \in \mathcal{F}$ (B1), (B2), and (B3) hold for $w \circ V$ and all the results of Chaps. 3 and 5 hold.

References

1. Anderson BDO, Moore JB (1971) Linear optimal control. Prentice-Hall, Englewood Cliffs, NJ.
2. Arkin VI, Evstigneev IV (1987) Stochastic models of control and economic dynamics. Academic Press, London.
3. Aseev SM, Kryazhimskiy AV (2004) The Pontryagin Maximum principle and transversality conditions for a class of optimal control problems with infinite time horizons, *SIAM J. Control Optim.*: 43, 1094–1119.
4. Aseev SM, Veliov VM (2012) Maximum principle for infinite-horizon optimal control problems with dominating discount, *Dynamics Contin. Discrete Impuls. Syst., Series B*: 19, 43–63.
5. Atsumi H (1965) Neoclassical growth and the efficient program of capital accumulation, *Review of Economic Studies*: 32, 127–136.
6. Aubry S, Le Daeron PY (1983) The discrete Frenkel-Kontorova model and its extensions I, *Physica D*: 8, 381–422.
7. Bachir M, Blot J (2015) Infinite dimensional infinite-horizon Pontryagin principles for discrete-time problems, *Set-Valued and Variational Analysis*: 23, 43–54.
8. Bachir M, Blot J (2017) Infinite dimensional multipliers and Pontryagin principles for discrete-time problems, *Pure and Applied Functional Analysis*: 2, 411–426.
9. Baumeister J, Leitao A, Silva GN (2007) On the value function for nonautonomous optimal control problem with infinite horizon, *Syst. Control Lett.*: 56, 188–196.
10. Blot J (2009) Infinite-horizon Pontryagin principles without invertibility, *J. Nonlinear Convex Anal.*: 10, 177–189.
11. Blot J, Cartigny P (2000) Optimality in infinite-horizon variational problems under sign conditions, *J. Optim. Theory Appl.*: 106, 411–419.
12. Blot J, Hayek N (2000) Sufficient conditions for infinite-horizon calculus of variations problems, *ESAIM Control Optim. Calc. Var.*: 5, 279–292.
13. Blot J, Hayek N (2014) Infinite-horizon optimal control in the discrete-time framework. SpringerBriefs in Optimization, New York.
14. Brock WA (1970) On existence of weakly maximal programmes in a multi-sector economy, *Review of Economic Studies*: 37, 275–280.
15. Carlson DA (1990) The existence of catching-up optimal solutions for a class of infinite horizon optimal control problems with time delay, *SIAM Journal on Control and Optimization*: 28, 402–422.
16. Carlson DA, Haurie A, Leizarowitz A (1991) Infinite horizon optimal control. Berlin, Springer-Verlag.

© The Author(s), under exclusive license to Springer Nature Switzerland AG 2019

A. J. Zaslavski, *Optimal Control Problems Arising in Forest Management*,

SpringerBriefs in Optimization, https://doi.org/10.1007/978-3-030-23587-1

17. Carlson DA, Jabrane A, Haurie A (1987) Existence of overtaking solutions to infinite dimensional control problems on unbounded time intervals, *SIAM Journal on Control and Optimizaton*: 25, 517–1541.
18. Cartigny P, Michel P (2003) On a sufficient transversality condition for infinite horizon optimal control problems, *Automatica J. IFAC*: 39, 1007–1010.
19. Coleman BD, Marcus M, Mizel VJ (1992) On the thermodynamics of periodic phases, *Arch. Rational Mech. Anal.*: 117, 321–347.
20. Cominetti R, Piazza A (2009) Asymptotic convergence of optimal harvesting policies for a multiple species forest, *Mathematics of Operations Research*: 34, 576–593.
21. Damm T, Grune L, Stieler M, Worthmann K (2014) An exponential turnpike theorem for dissipative discrete time optimal control problems, *SIAM Journal on Control and Optimization*: 52, 1935–1957.
22. De Oliveira VA, Silva GN (2009) Optimality conditions for infinite horizon control problems with state constraints, *Nonlinear Analysis*: 71, 1788–1795.
23. Evstigneev IV, Flam SD (1998) Rapid growth paths in multivalued dynamical systems generated by homogeneous convex stochastic operators, *Set-Valued Anal.*: 6, 61–81.
24. Gaitsgory V, Grune L, Thatcher N (2015) Stabilization with discounted optimal control, *Systems and Control Letters*: 82, 91–98.
25. Gaitsgory V, Mammadov M, Manic L (2017) On stability under perturbations of long-run average optimal control problems, *Pure and Applied Functional Analysis*: 2, 461–476.
26. Gaitsgory V, Parkinson, A, Shvartsman, I (2017) Linear programming formulations of deterministic infinite horizon optimal control problems in discrete time, *Discrete Contin. Dyn. Syst. Ser. B*: 22, 3821–3838.
27. Gaitsgory V, Rossomakhine S, Thatcher N (2012) Approximate solution of the HJB inequality related to the infinite horizon optimal control problem with discounting, *Dynamics Contin. Discrete Impuls. Syst., Series B*: 19, 65–92.
28. Gale D (1967) On optimal development in a multi-sector economy, *Rev. Econ. Stud.*: 34, 1–18.
29. Glizer VY, Kelis O (2017) Upper value of a singular infinite horizon zero-sum linear-quadratic differential game, *Pure and Applied Functional Analysis*: 2, 511–534.
30. Grune L, Guglielmi R (2018) Turnpike properties and strict dissipativity for discrete time linear quadratic optimal control problems, *SIAM J. Control Optim.*: 56, 1282–1302.
31. Gugat M, Trelat E, Zuazua, E (2016) Optimal Neumann control for the 1D wave equation: finite horizon, infinite horizon, boundary tracking terms and the turnpike property, *Systems Control Lett.* 90, 61–70.
32. Guo X, Hernandez-Lerma O (2005) Zero-sum continuous-time Markov games with unbounded transition and discounted payoff rates, *Bernoulli*: 11, 1009–1029.
33. Hayek (2011) Infinite horizon multiobjective optimal control problems in the discrete time case, *Optimization*: 60, 509–529.
34. Hammond PJ (1974) Consistent planning and intertemporal welfare economics. University of Cambridge, Cambridge.
35. Hammond PJ (1975) Agreeable plans with many capital goods, *Rev. Econ. Stud.*: 42, 1–14.
36. Hammond PJ, Mirrlees JA (1973) Agreeable plans, Models of economic growth. Wiley, New York, 283–299.
37. Jasso-Fuentes H, Hernandez-Lerma O (2008) Characterizations of overtaking optimality for controlled diffusion processes, *Appl. Math. Optim.*: 57, 349–369.
38. Jerry M, Rapaport A, Cartigny P (2010) Can protected areas potentially enlarge viability domains for harvesting management? *Nonlinear Anal. Real World Appl.*: 11 720–734.
39. Khan MA, Piazza A (2011) The economics of forestry and a set-valued turnpike of the classical type, *Nonlinear Anal.*: 74, 171–181.
40. Khan MA, Piazza A (2012) On the Mitra-Wan forestry model: a unified analysis, *J. Econom. Theory*: 147, 230–260.
41. Khlopin DV (2017) On Lipschitz continuity of value functions for infinite horizon problem, *Pure and Applied Functional Analysis*: 2, 535–552.

42. Kolokoltsov V, and Yang W (2012) The turnpike theorems for Markov games, *Dynamic Games and Applications*: 2, 294–312.
43. Leizarowitz A (1985) Infinite horizon autonomous systems with unbounded cost, *Appl. Math. and Opt.*: 13, 19–43.
44. Leizarowitz A (1986) Tracking nonperiodic trajectories with the overtaking criterion, *Appl. Math. and Opt.*: 14, 155–171.
45. Leizarowitz A, Mizel VJ (1989) One dimensional infinite horizon variational problems arising in continuum mechanics, *Arch. Rational Mech. Anal.*:106, 161–194.
46. Lykina V, Pickenhain S, Wagner M (2008) Different interpretations of the improper integral objective in an infinite horizon control problem, *J. Math. Anal. Appl.*: 340, 498–510.
47. Makarov VL, Rubinov AM (1977) Mathematical theory of economic dynamics and equilibria. Springer-Verlag, New York.
48. Malinowska AB, Martins N, Torres DFM (2011) Transversality conditions for infinite horizon variational problems on time scales, *Optimization Lett.*: 5, 41–53.
49. Mammadov M (2014) Turnpike theorem for an infinite horizon optimal control problem with time delay, *SIAM Journal on Control and Optimization*: 52, 420–438.
50. Marcus M, Zaslavski AJ (1999) On a class of second order variational problems with constraints, *Israel J. Math.*: 111, 1–28.
51. Marcus M, Zaslavski AJ (1999) The structure of extremals of a class of second order variational problems, *Ann. Inst. H. Poincaré, Anal. non linéaire*: 16, 593–629.
52. Maruyama T (1981) Optimal economic growth with infinite planning time horizon, *Proc. Japan Acad. Ser. A Math. Sci.*: 57, 469–472.
53. McKenzie LW (1976) Turnpike theory, *Econometrica*: 44, 841–866.
54. Mitra T, Wan HW (1985) Some theoretical results on the economics of forestry, *Review of Economics Studies*: 52, 263–282.
55. Mitra T, Wan HW (1986) On the Faustmann solution to the forest management problem, *Journal of Economic Theory*: 40, 229–249.
56. Mordukhovich BS (1990) Minimax design for a class of distributed parameter systems, *Automat. Remote Control*: 50, 1333–1340.
57. Mordukhovich BS (2011) Optimal control and feedback design of state-constrained parabolic systems in uncertainly conditions, *Appl. Analysis*: 90, 1075–1109.
58. Mordukhovich BS, Shvartsman I (2004). Optimization and feedback control of constrained parabolic systems under uncertain perturbations, Optimal control, stabilization and nonsmooth analysis. Lecture Notes Control Inform. Sci., Springer, 121–132.
59. Moser J (1986) Minimal solutions of variational problems on a torus, *Ann. Inst. H. Poincaré, Anal. non linéaire*: 3, 229–272.
60. Ocana Anaya E, Cartigny P, Loisel P (2009) Singular infinite horizon calculus of variations. Applications to fisheries management, *J. Nonlinear Convex Anal.*: 10, 157–176.
61. Piazza A (2009) The optimal harvesting problem with a land market: a characterization of the asymptotic convergence, *Economic Theory*: 40, 113–138.
62. Pickenhain S, Lykina V, Wagner M (2008) On the lower semicontinuity of functionals involving Lebesgue or improper Riemann integrals in infinite horizon optimal control problems, *Control Cybernet.*: 37, 451–468.
63. Rapaport A, Sraidi S, Terreaux JP (2003) Optimality of greedy and sustainable policies in the management of renewable resources, *Optimal Control Application and Methods*: 24, 23–44.
64. Rockafellar RT (1970) Convex analysis, Princeton University Press, Princeton, NJ.
65. Rubinov AM (1984) Economic dynamics, *J. Soviet Math.*: 26, 1975–2012.
66. Sagara N (2018) Recursive variational problems in nonreflexive Banach spaces with an infinite horizon: an existence result, *Discrete Contin. Dyn. Syst. Ser. S*: 11, 1219–1232.
67. Samuelson PA (1965) A catenary turnpike theorem involving consumption and the golden rule, *Amer. Econom. Rev.*: 55, 486–496.
68. Samuelson PA (1976) Economics of forestry in an evolving society, *Economic Inquiry*: XIV, 466–492.

69. Trelat E, Zhang C, Zuazua E (2018) Optimal shape design for 2D heat equations in large time, *Pure and Applied Functional Analysis*: 3, 255–269.
70. von Weizsacker CC (1965) Existence of optimal programs of accumulation for an infinite horizon, *Rev. Econ. Studies*: 32, 85–104.
71. Zaslavski AJ (1987) Ground states in Frenkel-Kontorova model, *Math. USSR Izvestiya*: 29, 323–354.
72. Zaslavski AJ (1995) Optimal programs on infinite horizon 1, *SIAM J. Control Optim.*: 33, 1643–1660.
73. Zaslavski AJ (1995) Optimal programs on infinite horizon 2, *SIAM J. Control and Optim.*: 33, 1661–1686.
74. Zaslavski AJ (1999) Turnpike property for dynamic discrete time zero-sum games, *Abstract and Applied Analysis*: 4, 21–48.
75. Zaslavski AJ (2005) The turnpike property of discrete-time control problems arising in economic dynamics, *Discrete and Continuous Dynamical Systems, B*: 5, 2005, 861–880.
76. Zaslavski AJ (2006) Turnpike properties in the calculus of variations and optimal control. Springer, New York.
77. Zaslavski AJ (2007) Turnpike results for a discrete-time optimal control systems arising in economic dynamics, *Nonlinear Anal.*: 67, 2024–2049.
78. Zaslavski AJ (2009) Two turnpike results for a discrete-time optimal control systems, *Nonlinear Anal.*: 71, 902–909.
79. Zaslavski AJ (2010) Stability of a turnpike phenomenon for a discrete-time optimal control systems, *Journal of Optimization theory and Applications*: 145, 597–612.
80. Zaslavski AJ (2010) On a class of infinite horizon optimal control problems, *Communications in Mathematical Analysis*: 9, 1–11.
81. Zaslavski AJ (2010) Good programs for a class of infinite horizon optimal control problems, *Communications on Applied Nonlinear Analysis*: 17, 1–14.
82. Zaslavski AJ (2011) The existence and structure of approximate solutions of dynamic discrete time zero-sum games, *Journal of Nonlinear and Convex Analysis*: 12, 49–68.
83. Zaslavski AJ (2011) Turnpike properties of solutions for a class of optimal control problems with applications to a forest management problem, *Dynamics of Continuous, Discrete and Impulsive Systems, Ser. B Appl. Algorithms*: 18, 399–434.
84. Zaslavski AJ (2014) Turnpike phenomenon and infinite horizon optimal control. Springer Optimization and Its Applications, New York.
85. Zaslavski AJ (2016) Structure of solutions of optimal control problems on large intervals: a survey of recent results. *Pure and Applied Functional Analysis*: 1, 123–158.
86. Zaslavski AJ (2017) Discrete-Time optimal control and games on large intervals. Springer Optimization and Its Applications, Springer, 2017.
87. Zaslavski AJ, Leizarowitz A (1997) Optimal solutions of linear control systems with nonperiodic integrands, *Math. Op. Res.*: 22, 726–746.

Index

Printed in the United States
By Bookmasters